Directors and the New Musical Drama

PALGRAVE STUDIES IN THEATRE AND PERFORMANCE HISTORY is a series devoted to the best of theatre/performance scholarship currently available, accessible and free of jargon. It strives to include a wide range of topics, from the more traditional to those performance forms that in recent years have helped broaden the understanding of what theatre as a category might include (from variety forms as diverse as the circus and burlesque to street buskers, stage magic, and musical theatre, among many others). Although historical, critical, or analytical studies are of special interest, more theoretical projects, if not the dominant thrust of a study, but utilized as important underpinning or as a historiographical or analytical method of exploration, are also of interest. Textual studies of drama or other types of less traditional performance texts are also germane to the series if placed in their cultural, historical, social, or political and economic context. There is no geographical focus for this series and works of excellence of a diverse and international nature, including comparative studies, are sought.

The editor of the series is Don B. Wilmeth (EMERITUS, Brown University), Ph.D., University of Illinois, who brings to the series over a dozen years as editor of a book series on American theatre and drama, in addition to his own extensive experience as an editor of books and journals. He is the author of several award-winning books and has received numerous career achievement awards, including one for sustained excellence in editing from the Association for Theatre in Higher Education.

Also in the series:

Undressed for Success by Brenda Foley
Theatre, Performance, and the Historical Avant-garde by Günter Berghaus
Theatre, Politics, and Markets in Fin-de-Siècle Paris by Sally Charnow
Ghosts of Theatre and Cinema in the Brain by Mark Pizzato
Moscow Theatres for Young People by Manon van de Water
Absence and Memory in Colonial American Theatre by Odai Johnson
Vaudeville Wars: How the Keith-Albee and Orpheum Circuits Controlled the Big-Time and Its Performers by Arthur Frank Wertheim
Performance and Femininity in Eighteenth-Century German Women's Writing by Wendy Arons
Operatic China: Staging Chinese Identity across the Pacific by Daphne P. Lei
Transatlantic Stage Stars in Vaudeville and Variety: Celebrity Turns by Leigh Woods
Interrogating America through Theatre and Performance edited by William W. Demastes and Iris Smith Fischer
Plays in American Periodicals, 1890–1918 by Susan Harris Smith
Representation and Identity from Versailles to the Present: The Performing Subject by Alan Sikes
Directors and the New Musical Drama: British and American Musical Theatre in the 1980s and 90s by Miranda Lundskaer-Nielsen

Directors and the New Musical Drama

British and American Musical Theatre in the 1980s and 90s

Miranda Lundskaer-Nielsen

First published in 2008 by
PALGRAVE MACMILLAN™
175 Fifth Avenue, New York, N.Y. 10010 and
Houndmills, Basingstoke, Hampshire, England RG21 6XS
Companies and representatives throughout the world.

PALGRAVE MACMILLAN is the global academic imprint of the Palgrave Macmillan division of St. Martin's Press, LLC and of Palgrave Macmillan Ltd. Macmillan® is a registered trademark in the United States, United Kingdom and other countries. Palgrave is a registered trademark in the European Union and other countries.

ISBN-13: 978–0–230–60129–1
ISBN-10: 0–230–60129–4

Library of Congress Cataloging-in-Publication Data is available from the Library of Congress.

A catalogue record for this book is available from the British Library.

Design by Newgen Imaging Systems (P) Ltd., Chennai, India.

First edition: April 2008

10 9 8 7 6 5 4 3 2 1

Printed in the United States of America.

Contents �◊᷵

List of Illustrations ⟡

Acknowledgments ✑

The research for this book was made possible by a Fellowship from the Graduate School of Arts and Sciences at Columbia University. In particular, I would like to thank Joy Hayton, the wonderful administrator of the English Department, as well as Professor Martin Meisel, Professor Martin Puchner, and especially Professor Arnold Aronson who gave invaluable feedback during the writing process. Laurence Maslon at New York University first gave me the idea for this book and his feedback and seemingly infinite knowledge of American musical theatre has been both a resource and an inspiration along the way. I am also indebted to Anni Parker, a fellow Anglo-American theatregoer, for her feedback and suggestions.

In addition to the Columbia University libraries, I made extensive use of the theatre archives of the V&A Theatre Collections in London and the Performing Arts branch of the New York Public Library and would like to thank all the helpful and efficient staff.

At Palgrave Macmillan, my thanks to Farideh Koohi-Kamali, Julia Cohen, and to my editor Don B. Wilmeth for their support, assistance, and encouragement throughout the editorial and production process.

My research for this book over the past few years has included conversations with British and American theatre professionals who gave generously of their time to answer my questions. Some of these interviews have been cited within the chapters and a selection of the transcripts is reproduced in full at the back of the book. I should note that the views and arguments expressed in this book are my own and are not necessarily all shared by my interviewees who represent a wide cross-section of backgrounds and perspectives. However, all these people have helped me to gain an understanding of the period that I cover and I am enormously grateful to them for sharing their professional experiences and insights with me. In particular, I would like to extend my thanks to writers and directors John Caird, William Finn, Adam Guettel, James Lapine, David Leveaux, Richard

Maltby, Jr., and Matthew Warchus; to Ted Chapin, president of the Rodgers and Hammerstein Organization, and Henry Little at the Arts Council of England; and to producers Marty Bell, Joanne Benjamin, Georgina Bexon, Chris Grady, Philip Hedley, Paul James, Margo Lion, Michael Lynas, Desi Moynihan, Tim Sanford, Deborah Sathe, Rachel Tackley, Jack Viertel, and Ira Weitzman.

On a personal note, I would like to acknowledge my brother Patrick Nielsen, an actor and director whose shared interest in theatre has been a source of great pleasure to me over the years. My parents, Jean and Tom Lundskær-Nielsen, also deserve a special mention. My interest in theatre stems from my early exposure to different kinds of performance and I am deeply indebted to them for giving me this wonderful gift.

Finally, I extend love and gratitude to my husband Michael Kenyon for his support, encouragement, and perceptive feedback throughout the preparation and writing of this book.

Introduction: Anglo-American Perspectives ∾

In the 1980s and 90s, a series of bold, commercially successful musicals arrived on Broadway from London's West End, with *Evita* followed by *Cats*, *Les Misérables*, *Me and My Girl*, *Phantom of the Opera*, *Starlight Express*, *Miss Saigon*, *Aspects of Love*, and *Sunset Boulevard*. This created a dilemma for the New York theatre industry and critics as well as for traditional Broadway theatregoers. There is no question that the shows arrived at a time when the Broadway musical was ailing, bringing in desperately needed audiences and revenue. Equally, *Cats* and its successors are widely acknowledged as having reinvigorated the "road"—the receiving houses outside New York that constituted the post-Broadway national tour. But the fact that the most successful shows on Broadway had originated abroad challenged the idea of musicals as an intrinsically American art form and the New York theatre community looked on in bemusement as the West End started to assume the role that Broadway had enjoyed in the previous four decades. To make matters worse, these shows did not look or sound like Broadway musicals and yet they were pulling in the kind of audiences that local producers could only dream of. Questions abounded. Where did these shows come from? Who were these artists? How did these musicals fit into the narrative of musical theatre history? And what did they signify for the future of the art form? In short, *what happened?*

This book highlights some of the cultural factors that underpin the extraordinary shifts in British and American musical theatre during the 1980s and 90s. Specifically, it offers an Anglo-American perspective on the way in which the recent developments have helped to broaden our perception of the musical as an art form, going beyond the usual focus on marketing, box office, and stage technology that has led to this being dubbed the era of the megamusical. By examining the critical reception, cultural contexts, and development processes of shows that came out of Broadway, the West End and the American nonprofit theatres in the 1980s and 90s, I highlight the evolution of new approaches to the art form and, specifically, the crucial role of directors in establishing a type of musical that I have called musical drama.

CRITICAL RECEPTION

One of the most distinctive factors of this era is the strength of the response from theatre critics and audiences. Interestingly, these have often been polar opposites. In New York, the imported West End musicals were enthusiastically embraced by local, national, and international audiences who kept the shows running for unprecedented lengths of time. From the "insiders" (the New York theatre industry, critics, historians, and traditional musical theatre fans), the response was less effusive and the tone and language employed in many theatre reviews at the time tell us as much about the critics' underlying assumptions about the art form as they do about the shows themselves.

In May 2001, opera critic Anthony Tommasini gave voice to a prevalent sentiment in the New York musical theatre community. In "They Do Write 'Em Like They Used To," a *New York Times* article about the annual Broadway TONY awards, Tommasini noted that while the nominees for Best Score had been "an embarrassment ever since Broadway was occupied by British invaders and Disney investors" the current contenders sounded reassuringly familiar: "What's refreshing, perhaps even a breakthrough in a backward-looking sort of way, is that the scores for 'The Producers' and 'The Full Monty' sound as if the last 25 years of pseudo-operatic spectacle had never happened. For that alone, each score deserves grateful cheers from die-hard devotees of the musical and awards galore next month."[1] He went on to suggest more explicitly that "one way to salvage the Broadway musical genre is to go back a couple of decades and start over."

While Tommasini is best known as a conservative opera critic, his comments express a common feeling of bemusement and resentment at the British musicals on Broadway—the inner sanctum of the American musical theatre. These feelings of hostility may explain the common reference to this period as the "British Invasion" with its connotations of the Revolutionary War and an unwelcome presence. It is not the first time this military phrase has been applied to cultural imports: notably, it was previously used to refer to the influx of British rock music to the United States in the 1960s. In the case of the musical, the phrase serves to reveal the intense American sense of ownership surrounding the art form that is so deeply embedded in American culture. The West End musicals met with a very mixed reception in New York ranging from rather grudging admiration (sometimes accompanied by an attempt to identify American elements or collaborators in the show) to disdain or downright hostility. Among these, *New York Times* critic Frank Rich wove an interesting path as one of the few critics to assess shows on their individual theatrical merits rather than rejecting them all out of hand. However, it is clear where his primary loyalties lay. In his 1987 article "The Empire Strikes Back," Rich gave voice to the dominant New York attitude to the West End musicals, framing the shows in commercial rather than artistic terms:

> For the New York theater, the rise of London as a musical-theater capital is as sobering a spectacle as the awakening of the Japanese automobile industry was for Detroit. Whether it is a real cultural phenomenon or merely a passing series of coincidences is another question. One could argue that the new London musical is a triumph of merchandising and of a handful of English artists, frequently abetted by Americans, rather than a significant and lasting artistic breakthrough.[2]

The reference to the American artists "abetting" the British creative teams is particularly telling here with its implication of enabling an act of wrongdoing. This attitude, found in contemporary journalism, theatre criticism, and theatre histories, has also pervaded the New York theatre industry. As the co-lyricist of *Miss Saigon*, musical theatre writer, director, and producer Richard Maltby Jr. was one of the few Americans to work on the West End shows. To many of his colleagues this amounted to cultural treason and he still describes himself wryly as "the Benedict Arnold of musical theatre."[3]

More alarmingly, perhaps, the existing scholarship on this period largely shares this Broadway-centric view. The pre-1980s musical has inspired a growing body of work that theorizes the musical through the lens of race, gender, ethnicity, sociocultural contexts, and gay and lesbian spectatorship. However, the current scholarship on the 1980s and 90s largely consists of general surveys overlaid with nostalgia, sporting elegiac titles such as *Broadway Babies Say Goodnight, The Rise and Fall of the Broadway Musical*, and *Ever After: The Last Years of Musical Theater and Beyond*. Theatre historians have overwhelmingly characterized this period as driven by commerce rather than artistry. What analysis there is of the shows themselves tends to focus almost exclusively on the perceived divergences from Broadway musical theatre traditions, such as the sung-through scores, new marketing strategies and technology behind musicals such as *Cats, Les Misérables, Phantom of the Opera*, and *Miss Saigon*. Where Broadway directors such as

Jerome Robbins, Michael Bennett, and Bob Fosse are (justly) acknowledged as central artistic figures within the musical theatre of the 1950s, 60s, and 70s, the directors of the 1980s and 90s have been largely sidelined in historical narratives in favor of the composers and producers.[4]

WHAT'S IN A NAME?

The common depiction of West End musicals as aberrations has led to some unhelpful terminology. In New York, the term "British musicals" has become common shorthand for large commercial musicals that place spectacle over artistry. This idea is furthered by the common adoption of the term "mega-musicals" to describe commercial West End and Broadway shows in the 1980s and 90s. The most comprehensive work in this respect is Jessica Sternfeld's *The Megamusical*. Sternfeld undertakes an illuminating and thoughtful analysis of many of the most successful commercial musicals of this era, offering a general production history of each before engaging in a more detailed musicological reading of shows that, as she puts it, "many scholars dismiss, disdain, and purposely ignore."[5] In her opening definitions, Sternberg notes that "the features that make a musical a megamusical come both from within the show itself and from its surrounding context" and goes on to lay out the general criteria:

> It features a grand plot from a historical era, high emotions, singing and music throughout, and impressive sets. It opens with massive publicity, which usually leads to millions of dollars in advance sales. Marketing strategies provide a recognizable logo or image, theme song, and catch phrase. Successful (re)productions spring up all over the world. Audiences rave; critics are less thrilled. It runs for years, perhaps decades, becoming a fixture of our cultural landscape.[6]

Using this definition as a guide, Sternberg goes on to discuss a broad range of shows that commonly fall under the megamusical banner, including *Jesus Christ Superstar*, *Cats*, *Les Misérables*, *The Phantom of the Opera*, *Blondel*, *Chess*, *Starlight Express*, *Aspects of Love*, *Sunset Boulevard*, *Miss Saigon*, *Jekyll and Hyde*, *Titanic*, *Beauty and the Beast*, *The Lion King*, *Rent*, *Ragtime*, *Aida*, and *The Producers*.

As Sternberg herself points out, classifying works of art is never an exact science and this is certainly true of an interdisciplinary art form like the musical. In his 1977 book on *Writing the Broadway Musical*, Aaron Frankel notes the list of terms in current usage including the "music drama" associated with Wagnerian opera as well as categories of Broadway musicals such as "revue," "musical comedy," "musical play," "Broadway opera," "new operetta," "plays with music," and "concept musicals."[7] Clearly, many of these are overlapping terms but crucially they are all defined by artistic criteria. By contrast, while Sternberg includes some artistic signifiers in her definition, the term "megamusical" in common usage is largely a reference to the external factors that she mentions rather than the tone, dramatic structure, or musical style that determine the other categories in Frankel's list. This creates two problems as we start to come to terms with recent musical theatre developments. First, "megamusical" is an inherently reductive term for an artistic work, implying that what is noteworthy about these shows is not their artistry but their size, be it big tunes, big scenery, big marketing campaigns, or big box office success. Second, the broad range of signifiers commonly used to identify megamusicals means that we end up grouping together shows that really have very little in common artistically. In particular, it obviates the need to look closer at individual shows in terms of the libretto and score, cultural origins, the relationship to the audience, or staging vocabularies—in short, the kind of artistic analysis that would reveal considerable differences between the shows that commonly fall under the "megamusical" banner.

Clearly, the recent past has seen a greater trend toward aggressive marketing, the stage technology has advanced, and there has been a shift toward through-sung scores and a different kind of subject matter than the Broadway musicals of the preceding decades. Equally, there are undeniable artistic difference between the West End musicals and their Broadway predecessors. The sense of wit and exuberance that marks the majority of Broadway musicals is less dominant in West End shows such as *Les Misérables* and *Miss Saigon*, and the tendency of the London musicals toward a through-scored and through-sung format has often sacrificed the linguistic nuances of the most sophisticated book musicals. There is also a different approach to set design, although this has often been a result of innovations in British theatre design and general technological advances rather than a simplistic love of spectacle. And the increased emphasis on the overarching theme is often at the expense of the star performances that are central to the Broadway musical theatre tradition.

However, rather than simply seeing the West End musicals as a failure to comply with the traditions of the Golden Age Broadway musical or seeing them as commercial ventures where size is everything, I would argue that they form part of a wider shift over the past few decades: namely the expansion of how we define the musical in relation to other performing arts. In 1991, American theatre critic and historian Martin Gottfried noted of the previous decade that "the British were not taking over our musical theatre really, although they were opening its eyes to possibilities for growth" and the result was "not that musical comedy or concept musicals had seen their last but that Broadway no longer could be monopolized with formula show-making by a small group of insiders."[8] In the 1980s and 90s, the commercially successful West End musicals found a way to reconnect with a broader audience and move the musical back toward the cultural mainstream. At the same time, the American nonprofit theatres began to embrace the musical theatre as an art form and to experiment with tone, form, and subject matter within the contexts of their particular artistic missions. Together, they started to create a more pluralistic idea of what musical theatre could look, sound, and feel like.

Given these developments, it seems to me that we might usefully look at the musical theatre of the 1980s and 90s as a number of different artistic impulses and try to develop a more precise vocabulary that reflects the artistry and cultural origins of individual shows. By looking beyond the traditional Broadway-centric narrative of musical theatre developments, we can start to see the 1980s and 90s as a time of broadening horizons, dramatic innovation, and formal experimentation in which the musical became a meeting point for a number of different theatre traditions.

MUSICAL DRAMA

If we look at the recent developments from a pluralistic rather than a Broadway-centric perspective—including British, American, commercial, and nonprofit musicals of this period—one of the most interesting "subgenre" to emerge is a group of shows that has as much basis in contemporary theatre as it does in the traditions of the Golden Age Broadway musical. I have called these shows "musical dramas."

At this point, a brief definition of terms might be helpful. The differences between the three most common musical theatre categories—musical comedy, musical plays, and concept musicals—have been discussed elsewhere and I do not intend to engage in detailed analysis here. However, for the benefit of newcomers to this subject, musical comedies are generally energetic, upbeat, and often witty in tone and include shows like *Anything Goes, Kiss Me, Kate, Annie Get Your Gun, Guys and Dolls*, and *Hello, Dolly!* Musical plays often tackle more serious subjects and place more emphasis on the "book" (the story, characters, and plot); the musical numbers are intended to advance and deepen the story rather than simply to entertain as is often the case with musical comedy. Early examples of musical plays include *Show Boat* (1927), which highlighted the human cost of misce-genation; *The Cradle Will Rock* (1937), a Brechtian-influenced musical written and set in the depression; *Pal Joey* (1940), which presented unusu-ally mature and unsentimental "romantic" leads; and *Lady in the Dark* (1941), which explored the new phenomenon of psychoanalysis. However, it was the writing team of Richard Rodgers (music) and Oscar Hammerstein II (book and lyrics) that is widely seen as establishing this genre of musical in the 1940s and 50s through shows such as *Oklahoma!* (1943), *Carousel* (1945), *South Pacific* (1949), and *The King and I* (1951) which combined memorable, melodious songs with psychologically moti-vated characters, thoughtful storylines, and serious social themes. The term "concept musical" was first applied to *Company* in 1970 and is now commonly used to describe musicals that are based around a central theme or idea.

The musical dramas that I look at combine the fundamental traditions of the Broadway musical plays and concept musicals (using songs written in a popular idiom to advance a story and/or explore a theme) with dramatur-gical and staging approaches from developments in nonmusical drama. By identifying this category of musicals, it becomes possible to draw artistic connections between ostensibly very different shows that originated in the 1960s and 70s on Broadway (*Cabaret, Company, Follies, Sweeney Todd*) and in the 1980s and 90s in the West End (*Evita, Les Misérables, Miss Saigon*) and in American nonprofit theatres (*March of the Falsettos, Falsettoland, Sunday in the Park with George, Jelly's Last Jam, Bring in Da Noise, Bring in Da Funk, Floyd Collins*).

My primary focus here is on the crucial role of director-dramaturgs and director-writers in the development of this body of work. There is, of course, a history of acknowledging musical theatre directors as central to the development of new work. The work of choreographer-director Jerome Robbins was key to this shift. In *West Side Story* (1957) and *Fiddler on the Roof* (1964), Robbins created a unified staging vocabulary to reflect the shows' themes and helped to raise the status of the director in the creative process. Subsequently, choreographer-directors such as Gower Champion, Bob Fosse, and Michael Bennett played a key role in shaping as well as stag-ing their shows. To call these artists *auteurs* would be inaccurate (the very collaborative nature of musical theatre makes this harder to achieve than in other performance arts) but certainly they were often what William Goldman calls "the muscle" in their shows, functioning as the most power-ful member of the creative team by shaping the development process and having the final say on creative matters.[9]

However, while these choreographer-directors—and later Broadway descendants such as Tommy Tune and Susan Stroman—have been accorded a central place in the historical narrative of musical theatre, the directors of musical dramas have so far been relatively ignored.[10] I would argue that this is an important omission and that in order to study these musicals, it is essential to recognize the vital contribution of a group of American and British theatre directors who have infused the mainstream musical with new ideas, techniques, and approaches. Starting with the work of Broadway producer and director Harold Prince in the 1960s and 70s, these include Trevor Nunn, John Caird, Nicholas Hytner, James Lapine, George C. Wolfe, Tina Landau, Sam Mendes, Matthew Warchus, and David Leveaux.

In particular, this group of directors effected two major changes in the developmental process for new musicals. First, they established a different kind of relationship between writers and directors. Unlike the Broadway directors of the 1960s and 70s who rose through the ranks as dancers and choreographers, the directors of musical drama came from text-based theatre. Their background in classical drama and new plays not only made them adept at dramaturgical work but also deeply respectful of the writers' intentions and instincts, prompting a collaborative rather than authoritarian approach. Second, while the more serious tone and subject matter of the musical dramas are foreshadowed in the Rodgers and Hammerstein shows, these new writers and directors took a more confrontational approach to social politics and to the theatre audiences themselves, often raising questions rather than resolving them, and eschewing the "feel-good" factor that characterize even the more daring Rodgers and Hammerstein musicals.

ANGLO-AMERICAN RELATIONS: AN HISTORICAL OVERVIEW

Given the impact of British musicals on Broadway in the 1980s and 90s, and the critical responses that have helped to shape our perception of the period, it seems clear that in order to understand these shows artistically it is necessary to understand the theatre traditions that they came out of and the Anglo-American cultural history that led to a sense of British musicals on Broadway being characterized as an "invasion."

While we commonly think of the musical as an intrinsically American art form, this was not always the case. Musical theatre is widely acknowledged to have developed out of eighteenth- and nineteenth-century European art forms such as the ballad opera, opéra comique, operetta, and

Gilbert and Sullivan's Savoy Operas as well as American vaudeville. In some ways the musical theatre landscape of the 1980s and 90s was a return to the first few decades of the twentieth century where it was a transatlantic art form. New York producer Florenz Ziegfeld frequently featured artists from Paris and London in his *Follies*; British actress Gertrude Lawrence played the lead in Broadway's *Oh Kay!* (1926);[11] and Noel Coward's *Bitter Sweet* (1929) was a huge success on both sides of the Atlantic combining a British author, an American star, and appropriation of European operetta traditions. Conversely, much of the work of the Guy Bolton /Jerome Kern/ P.G. Wodehouse writing team crossed the Atlantic to London while Broadway's Richard Rodgers and Lorenz Hart also wrote material for British impresario Charles Cochran's London revues.[12]

However, the term "musical theatre" has become synonymous with twentieth-century American culture and in particular the rich cultural heritage of New York's immigrant population. While the 1920s was a fertile time for the musical on Broadway, the supremacy of America (and specifically New York) as the musical theatre capital of the world can be dated to the "Golden Age" of the 1940s, 50s, and into the 60s.[13] It is true that there was a flurry of British musical imports in the 1960s, and John Degen has referred to this as a "British invasion" when (chiefly through the efforts of Broadway producer David Merrick) "musicals began to be imported wholesale to New York" including *Irma La Douce* (1960), *Stop the World, I Want to Get Off* (1962), *Oliver!* (1963), and the less successful transfers of *Oh, What a Lovely War!* (1964) and *The Roar of the Greasepaint, The Smell of the Crowd* (1965).[14] However, the term "invasion" should be viewed relatively: compared to the avalanche of Broadway musicals that transferred to London in these years, the flow in the other direction was more of a trickle. Certainly it did nothing to upset Broadway's preeminence as the world capital of musical theatre.

THE RISE OF THE BROADWAY MUSICAL

The enormous popularity of the musical in the Golden Age can in some measure be explained by the growing demand for accessible "culture" in this era. In *The Sixties: Years of Hope, Days of Rage*, Todd Gitlin notes that in the 1950s "middle-class Americans were becoming cultural omnivores."[15] This included increased travel and active participation in cultural activities both abroad and at home. As cultural historian David Savran has pointed out, cultural hierarchies became a national obsession and magazine articles began to appear in which different pastimes were ranked on a scale from highbrow to lowbrow.[16] The increase in cultural activity was reflected in the terminology that came into standard use: this was the era in which cultural hawks "no longer judged the opposition between highbrow and lowbrow sufficient to account for the unprecedented variety of cultural productions made in the United States, from bebop to *Queen for a Day*, from "Hound Dog" to *The Sound of Music*. . . ."[17] Increasingly, they began to recognize a third category of "middlebrow" culture to reflect the tastes of the growing ranks of middle-class cultural consumers.

This demand for middlebrow culture was a perfect fit for the developing Broadway Musical. The mixed heritage of the musical (drawing on opera, operetta, vaudeville, and drama) had always placed the art form at the populist end of the cultural spectrum, but in the 1940s and 50s it consolidated its status as middlebrow art with the rise of musical plays that were thoughtful adaptations of respectable source material and the influx of artists from more prestigious areas of the performing arts. In the 1940s, these included Lynn Riggs's *Green Grow the Lilacs* (*Oklahoma!*), Ferenc Molnar's play *Liliom* (*Carousel*), James Mitchener's book of short stories *Tales of the South Pacific* (*South Pacific*), Elmer Rice's *Street Scene* (*Street Scene*), Alan Paton's novel *Cry, the Beloved Country* (*Lost in the Stars*); in the 1950s, they included an adaptation of Bernard Shaw's play *Pygmalion* (*My Fair Lady*) and an updating of Shakespeare's *Romeo and Juliet* into a contemporary tragic love story (*West Side Story*). Mainstream Broadway musicals attracted artists such as composer and longtime Brecht collaborator Kurt Weill, African-American poet Langston Hughes, and classical composer and conductor Leonard Bernstein.

The advent of the original cast album in the 1940s became an important vehicle for the widespread dissemination of the musical, coinciding with the rise of consumer culture. Show music had been available to buy since the 1930s as recordings of hit songs by the stars of the shows. But following the success of *Oklahoma!* in 1943, record producers started to create original cast albums that included most or all of the songs.[18] These albums—sold alongside other kinds of pop music—helped to make the songs part of the popular culture by allowing for repeated listening and by bringing musicals into the lives and homes of people who had never set foot in a Broadway theatre. The popularity of the cast album with its uplifting show tunes might also help to explain the sunny image of even the darker musicals. Rodgers and Hammerstein shows such as *Oklahoma!*, *Carousel*, and *South Pacific* address some difficult social problems, but with a few notable exceptions (such as "Carefully Taught" in *South Pacific*) the uncomfortable ideas generally tend to be in the dialogue rather than the songs.

The success of the Golden Age Broadway musical can also be credited to the rise of television in the 1950s. While radio had previously enabled songs to travel beyond the theatre, television provided a more dramatic showcase. Of particular importance was the enormously popular *Ed Sullivan Show* that ran from 1948 to 1971.[19] This provided a showcase for new musicals as part of the eclectic programming that featured highbrow artists such as Rudolph Nureyev, Margot Fonteyn, Maria Callas, and classical pianist Eugene List alongside comedians such as Dean Martin and Jerry Lewis and the rising stars of the rock 'n' roll generation: Elvis Presley appeared on the show as did The Beatles, The Rolling Stones, The Doors, Janis Joplin, and Marvin Gaye. The Broadway musical was an ideal fit for this and many major stars appeared on the show to promote their shows, including Julie Andrews and Richard Burton in a scene from *Camelot*, Sammy Davis Jr. with the *Golden Boy* cast, and Yul Brynner performing material from *The King and I*.

The rise of the musical in America can also be explained in more sociopolitical terms. Broadway's Golden Age coincided with a period of national growth—economically, socially, and demographically—following World War II when the United States established itself as a powerful world leader. The energy, confidence, and social engagement of these postwar years were reflected in the Broadway musical and there was a noticeable increase in musical plays that reflected the dynamic social debates of the time. In *Our Musicals, Ourselves*, John Bush Jones explicitly makes this connection in his readings of the American musical as a reflection of wider social preoccupations. In particular, he identifies three major sociological shifts during the 1940s, 50s, and early 60s: the surge in prosperity after the war years; domestic social reform, including civil rights and race relations; and the nationalism bordering on xenophobia that followed World War II into the Cold War era. Thus Jones identifies in *Oklahoma!* and *Carousel* themes of national unification and criticism of blinkered entrepreneurship; he interprets *Brigadoon* and *Camelot* as shows about "isolationist Utopias"; he highlights the social liberalism in *Bloomer Girl* and *Finian's Rainbow*, one advocating women's rights and the other a satirical critique of bigotry and intolerance; and he emphasizes the theme of racial tension in *Lost in the Stars* and *West Side Story*.[20]

BRITISH MUSICALS DURING BROADWAY'S GOLDEN AGE

The sociological explanation for the rise of the Broadway musical can usefully be reversed to understand the demise of the British musical. While musical comedy in London started to wane even before World War I—becoming increasingly nostalgic rather than modern in tone—it was in the postwar years that the British musical lost its power as a cultural force when the optimism and sense of solidarity that characterized musical comedy began to seem irretrievably naïve in the face of a changing social climate. In *Musical Comedy on the West End Stage, 1890–1939*, Len Platt explains this decline partly in sociological terms, pointing out that the middle classes felt increasingly under threat from the mid-1920s to the late 1930s as trade unions gained a national political platform and the middle classes became radicalized in response.[21] These sociopolitical changes, coupled with the cultural shifts embodied in the new music, literature, and art in the 1920s and 30s, meant that the West End musical comedy was no longer in tune with the cultural shifts of the modern world.

In the following decades, the social and political shifts in Britain did little to alleviate these problems. Where the late 1940s and 50s saw America rising as a world power, postwar Britain was an empire in decline: the handover of India in 1947 was followed by the erosion of territories in Asia and Africa. As the nation tried to rebuild after the social and economic devastation of the war years, political focus shifted from empirical matters to domestic policy and the establishment of the welfare state in the mid- to late 1940s with the introduction of the National Health Service and the nationalization of coal, railways, and electricity. Economically, the British postwar recovery was built on a crippling loan from the United States and in 1949, in a grimly symbolic act of reversal, Britain had to acknowledge its increasing reliance on the former colony and accept a heavily devalued pound against the dollar. This economic and political relationship was echoed in the arts as American plays, music, and films started to infiltrate British popular culture. As Richard Eyre and Nicholas Wright point out, "The American presence—and the 'special relationship'—started to saturate British life politically, economically and culturally, invoking unease about our lack of democracy, unsatisfied desire for consumer goods, and restlessness and insecurity about our own culture. We became willingly, enthusiastically, and comprehensively colonized."[22]

From a practical standpoint, the lack of exciting and accomplished British musicals can be attributed to a dearth of committed creative producers in the West End. Musical theatre scholar John Snelson has argued persuasively that the perception of postwar Britain as a barren ground for musicals is partly due to scholars and critics advancing a "Broadway-led agenda which has denied [the] British shows their own home character."[23] As evidence, he cites shows such as *Bless the Bride* (1947), *Her Excellency*

(1949), and *King's Rhapsody* (1949) whose topicality to postwar Britain made direct comparisons to the American shows misleading. But the fact remains that Britain did not enjoy a musical theatre boom comparable to the one on Broadway, and this was to a great extent due to a lack of infrastructure. Snelson himself acknowledges that in the 1950s "the production of British musicals collapsed, principally because of the increasing lack of confidence of West End producers who preferred proven Broadway shows to the financial risks of unknown British ones."[24] In 1961, Noël Coward summarized the problem in his explanation of why he was premiering his new musical *Sail Away* across the Atlantic: "In New York light music is treated properly, taken seriously, and given a whole production expertise backstage which is still totally unknown in London. We just don't have the choreographers, the orchestrators, the dancers, the technicians to cope with the complexities of a big musical."[25]

Another reason for the lack of a British Golden Age in the 1940s and 50s is that the musical was perceived as commercial entertainment at a time when the British theatre was moving in a very different direction. The establishment of the Arts Council of Great Britain in 1946 (and with it the start of government funding for the arts) changed the landscape of British theatre with the emergence of subsidized companies such as the English Stage Company (1955), the Royal Shakespeare Company (1960), and the Royal National Theatre (1963), as well as prestigious regional theatres. (The English Stage Company soon became better known by its home at the Royal Court Theatre.) During the latter half of Broadway's Golden Age, Britain's most exciting writers, directors, and producers were found in the subsidized theatre, which was more influenced by the arrival of Brecht's Berliner Ensemble production of *Mother Courage* in 1956 than by the possibilities of musical theatre. Where the buoyant national mood in America found expression through musical theatre, the British postwar mood was more accurately reflected in the bleaker worlds of playwrights such as John Osborne, Arnold Wesker, Harold Pinter, and Edward Bond.

A NEW ERA

By the mid-1960s, however, the Golden Age of the Broadway musical was in decline, largely due to two major sociocultural developments. Mirroring city-dwellers around the country, the traditional Broadway audience (white, middle class) moved from Manhattan to the suburbs taking their disposable income and theatre-going habits with them: in the 1960s alone some

900,000 whites moved out of New York and were largely replaced by African Americans and Latinos with less cultural attachment to Broadway.[26] The other key factor in Broadway's demise was its inability to stay abreast with changing musical tastes and the newly politicized culture. As popular tastes shifted from showtunes to the sound of the Beatles, the Rolling Stones, the Doors, and Bob Dylan—and as an increasingly disillusioned generation started to question, criticize, and protest the status quo— Broadway musicals were left behind. All of these factors conspired to erode the Broadway audience base and weaken the position of musical theatre at the center of mainstream culture.

There were a number of different responses to this. One was to keep writing in the style of the Golden Age musicals with an increasing emphasis on nostalgia, such as the Jerry Herman musicals *Hello, Dolly!* (1964) and *Mame* (1966). Another was to reconceive the musical in terms of contemporary pop culture, as in *Hair* (1967) and *Jesus Christ Superstar* (1971). But a third response came from artists who started to approach the musical as contemporary drama and to explore ways in which the musical theatre might engage with audiences through different kinds of subject matter, musical style, characters, structure, and staging. And crucial to this cross-pollination was a group of theatre directors, starting with Broadway's Harold Prince, whose work on musical dramas helped to reposition the musical within the wider cultural landscape and open up new creative possibilities for one of our most popular art forms.

1. Harold Prince in Context ✤

Harold Prince is widely acknowledged as a crucial figure in the development of the postwar Broadway musical. He is associated with some of the most groundbreaking and acclaimed musicals of the 1950s, 60s, and 70s, first as a producer (*West Side Story, Fiddler on the Roof*) and then as producer-director (*Cabaret, Company, Follies, A Little Night Music, Pacific Overtures, Sweeney Todd*). In addition to his autobiography, *Contradictions*, Prince is the subject of several profiles, most notably Carol Ilson's *Harold Prince: A Director's Journey* and Foster Hirsch's *Harold Prince and the American Musical Theatre*, which focus on his dual identity as the inheritor of Broadway showmanship and as a pioneer in bringing darker themes to the Broadway musical theatre.

What has not been as widely recognized is how Prince's work foreshadowed the musical dramas of the 1980s and 90s by redefining the relationship between musical theatre and contemporary drama in two crucial ways. First, he used the musical to explore the same kind of social questions that were engaging social dramatists such as Neil Simon, Edward Albee, and David Rabe. Second, his musicals developed a more confrontational relationship with the audience through staging that drew on elements of Russian Expressionism, British Music Hall, French Grand Guignol, German Expressionist film, and on the work of practitioners such as Bertolt Brecht, Yuri Petrovich Lyubimov, Joan Littlewood, and film director Friedrich Wilhelm Murnau. By incorporating these influences, Prince moved the musical beyond the traditions of the American musical, widening the frames of reference and, in the process, establishing a new kind of musical drama.

In the following discussion, I do not wish to imply that Prince was an *auteur* in the sense of directors such as Robert Wilson or that he was the sole instigator and influence on these shows. Prince's work has always been collaborative, and one of his great gifts has been to surround himself with world-class writers and designers. Indeed, many of his shows are more commonly identified with his collaborators than with Prince, such as composer-lyricist team John Kander and Fred Ebb (*Cabaret*), Andrew Lloyd Webber and Tim Rice (*Evita*), and composer-lyricist Stephen Sondheim (*Company, Follies, A Little Night Music, Pacific Overtures, Sweeney Todd*). In addition, much of his staging owes a great debt to the vision, imagination, and ingenuity of designers such as Boris Aronson, Eugene Lee, and Maria Björnsson. It is not my intention to deny the contributions of these artists, which have been extensively documented elsewhere; my aim here is to explore the impact of Prince's approach to theatre on the development of musical drama on Broadway. From *Cabaret* onward, Prince performed multiple roles as producer, co-conceiver, dramaturg, and director that often made him the "muscle" in the creative process.[1] As the producer of his own work, Prince was in a unique position to hire collaborators who shared his vision and, although not a writer himself, he had a profound influence on choosing and shaping his directorial projects, establishing the tone and point of view, building characters, and honing the dramatic structure in such a way as to broaden the horizons of the Broadway musical and lay the groundwork for the directors of musical drama in the 1980s and 90s.

SOCIAL CONTEXTS

Prince spent his formative professional years on Broadway at a key moment in America's social and cultural development and the critical, questioning tone of his musicals reflect the era in which they were created. The 1960s have rightly become synonymous with sweeping social changes. Through their social politics, economics, fashion, music, and literature, the generation who came of age in this decade actively and loudly questioned the conformist culture of the 1950s, replacing it with a more individualistic, experimental, and spiritual approach to life. Social structures were redefined, with the traditional family unit often superceded by social groups such as the hippie "tribe" or organizations based on race, gender, and sexuality. The growing feminist movement questioned the idea of women's primary identity as wives and mothers, finding a voice through publications such as Betty Friedan's *The Feminine Mystique* (1963), Mary McCarthy's novel *The Group* (1963), and (in smaller circles) the distribution of Casey Hayden and Mary King's manifesto on "Feminism and the Civil Rights Movement" (1965). The sanctity of marriage was undermined by increasingly liberal ideas about sex, the introduction of the contraceptive pill in the early 1960s, and California's legalization of the no-fault divorce in 1969. That same year, a police raid on the gay and transsexual Stonewall Inn in New York served as a catalyst for the gay rights movement. This period also saw the intensification of the Civil Rights Movement, with hundreds of local uprisings, student sit-ins, and numerous high-profile marches: in August 1963, the March for Jobs and Freedom culminated in 250,000 people gathering at the Lincoln Memorial in Washington DC to hear Martin Luther King announcing "I have a dream . . . ."[2] Most prominently, perhaps, the 1960s were marked by increasingly vociferous public protests against the war in Vietnam. The national pride and confidence of the post–World War II era gave way to feelings of doubt, guilt, and a sense of betrayal.

CULTURAL CONTEXTS

These radical social shifts were echoed throughout the arts scene during the 1960s and 70s as artists began to conduct formal and thematic experiments in music, visual arts, and drama. Musically, the rock 'n' roll and Motown sounds of the 1950s and early 60s were replaced by the arrival of British rock bands including The Beatles and The Rolling Stones. Political singer-songwriters such as Bob Dylan became the voice of the more socially idealistic generation. The art world saw a radical shift, too, as the early trailblazing work of artists such as Jackson Pollock and Mark Rothko in the 1940s and 50s was followed by further formal experimentation in the 1960s, not least through the rise of pop art and artists such as Andy Warhol.

By the time Prince started directing in the mid-1960s, the American theatrical landscape had also changed significantly from the time of the Golden Age musicals, both in terms of infrastructure and aesthetic parameters. Most obviously, the geography of the American theatre expanded, with Off Broadway creating an alternative home for New York theatre and the Regional (or Resident) theatre movement starting to challenge the Broadway-centric model of generating theatre. These changes did not always result in radically experimental work. Since 1950, Off Broadway has been a technical term employed by Actors' Equity to denote Manhattan theatres with a particular seating capacity and a set union pay scale: then (as now) Off Broadway included commercial as well as more experimental productions. Equally, many regional theatres were simply trying to provide their own homegrown productions rather than importing them from Broadway.

However, the growth of the American theatre had significant aesthetic repercussions as many of the new theatres encouraged artists to explore different kinds of stories and storytelling. In New York, the 1950s saw the emergence of theatre companies like the Living Theatre (1948), Circle in the Square (1952), and the New York Shakespeare Festival (1954), all of them attempting new approaches either to classical plays, new writing, or both. This was echoed across the country with vibrant new theatres starting to create their own work, including Margo Jones's Theatre '47 in Dallas (1947), Nina Vance's Alley Theatre in Houston (1949), and Zelda Fichandler's Arena Stage in Washington DC (1951).[3] In 1965, the establishment of the National Endowment for the Arts resulted in even more new theatre companies and institutions.

MUSICALS IN THE 1960s AND 70s

The new developments in American drama were not generally reflected in musical theatre, although a few isolated musicals did begin to appear Off Broadway that started to open up new possibilities for the genre, expanding the kinds of subject matter and staging that could be explored through musical theatre. *Hair* famously brought the hippie movement to the musical stage while *Getting My Act Together and Taking It on the Road* did the same for feminism and *Boy Meets Boy* featured overtly gay characters. The most interesting figure in this respect is probably director Tom O'Horgan, whose artistic roots were in the highly experimental Off-Off-Broadway theatre La MaMa Etc. but who went on to direct the Broadway productions of *Hair* (1968) and *Jesus Christ Superstar* (1971). O'Horgan's staging was highly visual and structurally experimental, based to a great extent on a physical translation of texts and ideas into film, mime, music, and dance. His work on *Hair* retained these avant-garde impulses: specifically, Arnold Aronson has drawn connections between *Hair* and the neo-Expressionist physical ensemble work of the Living Theatre and the Open Theatre.[4] Altogether, the musical provided a new and more active experience for Broadway audiences, with O'Horgan making unexpected use of the auditorium and the stage space: as Gerald Berkowitz notes, "Singers appeared in the aisles, popped out of trap doors, climbed the walls and hung from the proscenium arch"—and, of course, took off their clothes in one highly publicized but dimly lit scene.[5]

Overall, however, Broadway continued to produce musical plays and comedies that followed the models established in the 1940s and 50s, demonstrating only a cursory interest in the social, cultural, and theatrical shifts outside the theatre doors. In "The Dream Shattered: America's Seventies Musicals," Barbara Means Fraser argues convincingly that many musicals of this period manifested a darker sensibility and a stronger measure of cynicism and disillusionment than the more wholesome worlds of Rodgers and Hammerstein—a reaction to an era where "the basic institutions that had stabilized our country in times of trouble were the very culprits of the disillusionment."[6] However, this is manifested much more strongly in the Prince-directed musical dramas that she mentions (*Sweeney Todd, Follies*) than the musicals staged by choreographer-directors Bob Fosse (*Chicago*) and Michael Bennett (*A Chorus Line*). Despite its credentials as a concept musical, *A Chorus Line* is closer to musical comedy than musical drama: any implicit criticism of the American dream in the chorus dancers' struggle is drowned out by the celebratory tone of the musical numbers and, tellingly, the most memorable song ("What I Did For Love") is in fact a love song to the theatre. Equally, *Chicago*—ostensibly a show about corruption and crime—dresses its cynicism in sequins and is fundamentally a satirical musical comedy that emphasizes two comic star turns, sleek production numbers, and Fosse's distinctive choreography of splayed hands, jazz isolations, and erotically charged gestures.

By contrast, Prince's early work as a producer-director-dramaturg on *Cabaret* (1966), *Company* (1970), and *Follies* (1971) addressed some of the most fundamental social issues of the day. The direct impetus for *Cabaret* was the intensifying Civil Rights movement and the show's creators saw a clear parallel between the racial prejudices that allowed the Nazis to gain power and contemporary racial prejudice in America.[7] Four years later, *Company* gave voice to the increasingly complicated attitude toward the institution of marriage as a result of the redefinition of gender roles by the growing feminist movement. (Produced the year after the Stonewall riots ignited the Gay Rights movement, the show also lends itself easily to a gay reading although the creative team has maintained that the central character of Bobby is a straight man who is afraid of commitment rather than a closet homosexual.) In 1971, *Follies* reflected the national mood following the American engagement in Vietnam. Not only was this a military defeat, but the sense that America was engaged in an immoral war led to a sense of a lost Eden—a feeling that was reflected in the musical's twin underlying themes of nostalgia and disillusionment.

PRINCE AND INTERWING THEATRE

Perhaps the most useful way to classify Prince's work in relation to the Broadway musical and the more experimental theatre of his day is through the term "interwing" theatre. In *A Theater Divided* (1969), Martin Gottfried identifies two kinds of theatre in mid-twentieth-century America, which he terms "left wing" (theatre that innovates and challenges) and "right wing" (theatre that replicates and reassures). Gottfried singles out the musical as the one kind of theatre that occasionally manages to straddle these opposite impulses to create "interwing" shows that combine the technical expertise of the right wing and the innovations of the left-wing theatre.[8] A closer examination of Prince's musicals reveals the "interwing" nature of his work: namely, his ability to combine the formal experimentation of the Off-Broadway and avant-garde theatre movements and the thematic preoccupations of contemporary social dramatists with the sophisticated stagecraft and structural precision of the Broadway musical.

Despite Prince's affiliation with Broadway, his approach to staging has clear parallels in the more avant-garde theatre of his day. Prince's use of direct audience address in *Cabaret* and his repeated use of the chorus as an extension of society (and of the audience) in shows such as *Cabaret*, *Company*, *Evita*, and *Sweeney Todd* echo the preoccupation of the avant-garde theatre with confronting the audience and making them complicit in the theatrical event. In addition, Prince's interest in formal experimentation can be seen in his Off-Broadway work, including an Expressionist production of Durrenmatt's provocative *The Visit* and his original, environmental staging of *Candide* in a 180-seat theatre at the Brooklyn Academy of Music with the audience seated on bleachers amongst runways and ramps.

However, while both these productions demonstrate Prince's genuine interest in exploring theatrical alternatives to Broadway, he is not an avant-garde artist in the mold of Richard Foreman and Robert Wilson and he retains a traditional approach to the supremacy of plot, character, and text within the theatrical event. It is no accident that Prince's main venture into Off-Broadway work was with the Phoenix Theatre—which was founded by experienced commercial producers and maintained some links to Broadway—rather than more radical theatres such as the Living Theatre or La MaMa ETC. Equally, his Off-Broadway production of *Candide*, while experimental, was not too avant-garde to transfer to Broadway.

OKLAHOMA! VERSUS *COMPANY*

The distance between Prince's musical dramas and the musical plays that preceded them can be seen in a comparison between Rodgers and Hammerstein's breakthrough musical *Oklahoma!* and the Prince musical *Company*.

In 1943, *Oklahoma!* signaled a new era for the Broadway musical, establishing the dominance of musical plays on Broadway. Based on Lynn Riggs's play *Green Grow the Lilacs, Oklahoma!* tells the story of farmers and cowboys during the taming of the West. The storyline centers around Laurey, the niece of farm owner Aunt Eller, who attempts to choose between the swaggering cowboy Curly and the brooding farmhand Jud. The show's book and musical numbers touch on broader social themes— the antagonism between farmers and cowhands, the importance of assimilation in a volatile society—and through a dance number called "Laurey Makes Her Mind Up" (originally choreographed by Agnes de Mille as a Freudian dream) it explores Laurey's fearful fascination with Jud in terms of the subconscious desires that govern sexual attraction.

Company is set in the late 1960s when it was created. While the show was written by librettist George Furth and composer-lyricist Stephen Sondheim, it was Prince who saw the potential for a thematic musical in Furth's collection of short plays on the theme of marriage and who acted as a dramaturg as well as director and producer. The show is structured as a series of vignettes around the central character of Bobby, a single 30-year-old in Manhattan whose life consists of casual romantic relationships and his friendships with married friends for whom he serves as a confidant. The show portrays Bobby's ambivalent response to settling down with one person, and the musical numbers highlight both the costs and benefits of marriage as seen through the eyes of Bobby and his social circle.

Company both honors and challenges the tradition of the Rodgers and Hammerstein musicals as embodied in *Oklahoma!* Like the earlier musical, *Company* is a product of the sociocultural context in which it was created. However, while *Oklahoma!* encompasses some serious questions such as social assimilation it largely reflects the sense of patriotism that was prevalent in 1943; by contrast, *Company* reveals the rather more complicated attitude toward social relationships after the upheavals of the 1960s. Where *Oklahoma!* reassures and celebrates, upholding marriage as the bond that unifies society, *Company* questions the institution of marriage and highlights the inevitable compromises that a union between two individuals entails. A comparison between the finales of the two shows clearly illustrates the difference between them. In Max Wilk's account of the creation of *Oklahoma!* he notes that after an encouraging first half on opening night, "on rolled that second act, gathering momentum as it went, resonating to an enraptured audience that laughed and sighed in all the right places" until "that exultant 11 o'clock chorale version of 'Oklahoma' with the cast down front belting out to the New Yorkers a hymn to a brand new state: 'Oklahoma, O . . . K!'"[9] The *Evening Post* critic noted approvingly that although the show had "a mild, somewhat monotonous beginning," it picked up steam and turned into a rousing, festive affair: "By the time Curly and Laurey were married and everybody was out shouting the title song 'Oklahoma!' we were having a very good time."[10] The ending of *Company* is much more muted. The original final song (a forceful rejection of marriage) was vetoed by Prince for being too negative; however, even in the modified version, the tone is hesitant and philosophical rather than jubilant. In "Being Alive" Bobby makes only a statement of *intent* to marry; moreover, it is couched in ambivalent terms that suggest a fear of loneliness and an intellectual embrace of the sacrifices and compromises of marriage rather than unbridled enthusiasm. Tellingly, his decision to marry is not connected to anyone in particular. Earlier in the show his friend Amy points out to him that it is not enough to love "*some*body"—rather, you have to love "some*body*." Bobby's decision at the end that he is ready to commit himself to a relationship is tempered by the fact that he is addressing his final song to an unknown "somebody" ("Somebody hold me too close / Somebody hurt me too deep").[11] Where *Oklahoma!* ends with a joyous wedding celebration and a full-cast reprisal of the triumphant title song, the final number in *Company* has Bobby alone onstage persuading himself that committing to another person might, on balance, be a good thing.

In terms of the Rodgers and Hammerstein musicals, *Company* is probably more closely descended from their most experimental (if commercially unsuccessful) musical. *Allegro* (1947) is an episodic allegory about a man who climbs the ladder of professional and economic success but loses himself in the process. Prince and librettist George Furth were both admirers of *Allegro* as was Hammerstein's protégé Stephen Sondheim who started his career as an assistant on the show. Sondheim later reflected that while much of the Rodgers and Hammerstein canon consists of realistic plays with songs, "in *Allegro* Oscar used theatre convention frankly as theatre convention: the chorus that oversees the action, the abstract scenic design. These techniques have had a large influence on Hal's work and on mine."[12] After *Allegro* failed at the box office, Rodgers and Hammerstein returned to more traditional territory and the musical's dramaturgical innovations became a digression rather than immediately forging a new path for the mainstream musical.

MENTORS AND PEERS

Prince's relationship to—and distance from—the dominant Broadway musicals can also be measured through a comparison with his directing mentors and peers. In his autobiography, *Contradictions*, and in numerous interviews he has acknowledged the influence of George Abbott and Jerome Robbins. However, this debt is one of stage technique rather than underlying philosophy regarding the musical's relationship to the audience and to the world around it. As the coproducer of *West Side Story* and producer of *Fiddler on the Roof*, Prince observed Robbins at first hand. Trained as a ballet choreographer, it was Robbins who introduced Prince to the idea of a unified production vocabulary and to a more fluid use of the stage. In his autobiography, Prince notes how Robbins used the entire depth of the stage rather than moving actors horizontally (as was traditional in musical theatre) and the importance he attached to the aesthetic use of lighting rather than the usual approach of "lights up for the scene, and lights down for the song."[13] He also recalls the effectiveness of improvisation and empathy techniques in Robbins's rehearsals—elements of "the method" that he had learned under Lee Strasberg at The Actors Studio. These included hanging articles about real gangs backstage at *West Side Story* and deliberately creating tension between the actors playing the Jets and the Sharks to allow them to access personal feelings akin to those of their characters. In 1966, this technique was used by Prince on *Cabaret* where he brought in a contemporary photograph of young white Americans at an anti-integration rally in the South to point up the parallels between anti-Semitism in Weimar Berlin and racism in 1960s America.[14] Prince's admiration for George Abbott was also based more on stage craft and discipline than on philosophy. As a producer, director, and bookwriter, Abbott specialized in energetic, light musical comedies: his directing credits include *Billion Dollar Baby* (1945), *High Button Shoes* (1947), *Wonderful Town* (1953), *How Now, Dow Jones* (1967) and his work as a book writer and director includes *Where's Charley* (1948) and *A Tree Grows in Brooklyn* (1951). His approach to directing, typical of Broadway musicals at the time, was a very pragmatic one and Prince came to emulate his mentor's discipline and professionalism.[15]

However, while Prince inherited staging and rehearsal techniques from his colleagues, his vision of theatre as a cultural force was very different, emphasizing the musical stage as a place to explore topical and uncomfortable issues. Jerome Robbins, Gower Champion, Bob Fosse, and Michael Bennett graduated from the chorus line to choreography and direction and although they reinvigorated the musical through new approaches to physical staging, their work remained largely unaffected by the social revolution that was occurring outside the theatre. Tellingly, as Bennett and Fosse gained more influence over their shows, they focused increasingly on musicals set in the world of show business where social commentary was secondary to distinctive staging: although touching on genuine social issues, Bennett's *A Chorus Line* and *Dreamgirls* and Fosse's *Sweet Charity* and *Chicago* are ultimately stories that celebrate individual resilience in the face of adversity. Years later, Bennett's longtime choreographic partner Bob Avian worked on *Miss Saigon* and *Martin Guerre* in London but noted of Bennett that "he would not have accepted these shows. He didn't do anything he didn't understand—like 16th century France. He wouldn't have even read the script. He was a kid from Buffalo, he didn't even finish high school—although he won his Pulitzer Prize."[16] And even though Robbins was instrumental in the development of musicals such as *West Side Story* and *Fiddler on the Roof,* his contribution did not consist of honing the social politics but of creating a unified production vocabulary through fluid staging.

By contrast, Prince's college years helped to foster his fascination for history, playwriting, and the world around him.[17] The subject matter and the sometimes confrontational tone of *Cabaret* and *Follies* can be traced to this key divergence from George Abbott as voiced by Prince in a TV interview: "I have a darker sensibility . . . I'm political, he's not. I'm issue-oriented, he's not."[18] This translated also into a very different attitude toward the audience. "He really unabashedly wants people to have a good time," Prince explained, "and sometimes I don't give a damn. I want to stimulate them." Where Abbott created shows to uplift and amuse his audiences, Prince aimed to use the emotional and seductive power of musicals to address serious subjects and make the audiences confront or at least acknowledge their own prejudices, fears, and insecurities.

PRINCE AND BRECHT

In a sense, the difference between Prince and his colleagues can be seen as the tension between the Wagnerian and Brechtian impulses of the Broadway musical. In *Making Americans: Jews and the Broadway Musical,* Andrea Most notes that there are points of similarity between the separation of different theatrical elements in Brecht's *verfremdungseffekt* ("alienation technique") and the presence of music, dance, and spoken dialogue in traditional musical comedy. Indeed, Mark Blitzstein's 1938 political musical *The Cradle Will Rock* was actually dedicated to Brecht. However, there is also a connection between the emotional and cohesive emphasis of the Wagnerian *gesamtkunstwerk* and the rise of the integrated musical, and Most notes that "when Broadway musical composers and writers began to refashion themselves as 'artists' in the postwar period by integrating songs with dialogue, they turned not to Brecht but to Wagner for their inspiration."[19] The musicals spearheaded by Bennett and Fosse, based around a central directorial vision, can in a sense be seen as examples of *gesamtkunstwerk*, although they still retained a level of showmanship and a relationship to the audience that revealed their roots in vaudeville and musical comedy.

Prince's desire to stimulate his audiences and to confront them with uncomfortable social issues connects him more strongly to the philosophy laid out in the seminal essay "The Modern Theatre is the Epic Theatre" in which Brecht contrasts the "epic theatre" with the "dramatic theatre." While not all of his criteria apply to Prince, his definition of epic theatre as something that "turns the spectator into an observer," "forces him to take decisions," makes him "face something," and brings him to the "point of recognition" is strongly applicable to almost all of Prince's work.[20] Prince himself has denied a conscious debt to Brecht, objecting mainly to the emphasis on "alienation" and arguing that "I want to be enveloped, engaged in the theatre."[21] The desire to "envelop" the audience both mentally and emotionally is evident in Prince's work, not only through empathy with characters and through seductive musical numbers but also through his use of the theatre space: in *Cabaret*, a large tilted mirror onstage reflected the auditorium from the audience's perspective and in *Evita* he used roving searchlights to create the sense of being at a political rally. However, there is a clear correlation between Brecht and Prince in their view of theatre as a forum for addressing serious contemporary social issues and their use of nonnaturalistic narrative techniques to forge a direct relationship with the audience.

This sociopolitical approach to musical theatre did not sit well with all audiences or critics. Both *Cabaret* and *Follies* had audiences walking out in sizeable numbers, objecting to the dark and uncomfortable subject matter and tone. *Cabaret* also came under fire from Jewish groups for one supposedly anti-Semitic lyric when the Emcee refers to his gorilla girl friend as Jewish in "If You Could See Her through My Eyes." The number was intended as a criticism of anti-Semitism in popular entertainment, but was not universally understood that way. It would be inaccurate to paint Prince as uninterested in popular appeal—as a commercial producer, this was part of his job and he did compromise at times, such as when he made the writers amend the offending lyric in *Cabaret* and demanded a less depressing ending for *Company*. However, it is also clear that overall his interest was in creating work that could engage with topical issues, even if it reduced the mass appeal of the show.

PRINCE AND CONTEMPORARY
AMERICAN DRAMA

In addition to the Brechtian impulses, a closer examination of *Company* and *Follies*—both set in contemporary New York City—reveals a thematic connection between Prince and some of the leading social dramatists of the day. In 2002, Stephen Sondheim reflected that "until about 30 years ago, people went to musicals not wanting to be anything but entertained on a very easy level. But starting in the late '60s and '70s, we began experimenting with form, with content. Musicals were trying to encroach on the territory that previously had belonged solely to plays."[22] Certainly, *Company* was an anomaly among the other musicals that opened in 1970. These included *Oh! Calcutta!* (a revue that featured onstage nudity), *Coco* (a biographical showcase for Katherine Hepburn as Coco Chanel), and *Applause* (a backstage musical showcasing a series of prominent actresses starting with Lauren Bacall).

Considered in the context of contemporary plays, however, *Company* is less unusual. The characters and the central theme of *Company* resonate with many of the plays that were appearing at a time when sexual taboos were loosening and the social and marital roles of both men and women were being redefined. On Broadway, Neil Simon had a stream of plays and musicals produced from the early 1960s, three of which dealt specifically with the malaise of married couples. Although he excelled in witty dialogue and comic punch lines, there is an underlying cynicism and sense of loss in plays such as *Plaza Suite* (1968), *Last of the Red-Hot Lovers* (1969), and *California Suite* (1976). In all three, husbands and wives struggle with varying degrees of success to juggle their personal needs with marriages that both define and strangle their sense of self. Even sharper was Edward Albee's *Who's Afraid of Virginia Woolf?* which, from 1962 to 1964, offered Broadway audiences a vicious dissection of middle-class married life. Nor was *Company* at the tail end of this trend: eight years after the musical opened, *Curse of the Starving Class* (1978) was the first of Sam Shepard's family plays—followed by *Buried Child* (1978), *True West* (1980), and *A Lie of the Mind* (1985)—all of which undermined the myth of the supportive family unit as the center of American life.

Certainly, there is a difference of tone and central preoccupations between *Company* and the plays listed earlier, but there are also telling similarities which indicate that they have all come out of the same period in American social history. *Company*'s cynical, wise-cracking, alcoholic society wife Joanne inhabits a world very similar to that of Neil Simon's *Plaza Suite*, while Bernie's middle-aged crisis in *Last of the Red-Hot Lovers* has clear connections to the identity crisis of the two couples in *Follies* as they measure the distance between their youthful selves and their current status as unfulfilled, lonely individuals married to the wrong people. Similarly, the married men in *Company* speak for many of the married couples in the plays, not least in songs such as "Sorry-Grateful" ("You're always sorry / You're always grateful / You're always wondering what might have been")[23] or "Have I Got a Girl For You." In the latter, Bobby's married male friends try to live vicariously through him, idealizing his unfettered, romantic single life and contrasting it with the prosaic reality of married life: "Whaddaya like—you like indescribable bliss? / Then whaddaya want to get married for?"[24] Bobby's genuinely mixed feelings about the marital state—although out of step with those of romantic musical comedy—are entirely in keeping with the frustrations and fears that underpin the Albee, Simon, and Shepard plays.

Another interesting if initially unlikely connection can be made between *Follies* (with music and lyrics by Stephen Sondheim, book by James Goldman, and co-conceived and directed by Prince) and David Rabe's play *Sticks and Bones*. Both shows premiered in 1971 (on Broadway and Off Broadway, respectively) and both used well-known icons from the past to emphasize the change of mood in post-Vietnam America. Ostensibly, *Follies* depicts a reunion of ex-Follies girls, focusing on the unhappy marriages of two of the women, Sally and Phyllis, and their husbands, Buddy and Ben. During the evening it transpires that all four are unfulfilled—Sally has always been in love with her idealized fantasy of Ben, Buddy has a mistress who loves him while he remains in love with Sally, Ben feels cheated by the emptiness of his financial success, and Phyllis's hollow existence as Ben's society wife has resulted in a combative and cynical world-view. As the ex-Follies girls reminisce around them, the two couples' anxieties and frustrations start to spill out, and the memories of their carefree, optimistic youth highlight the disillusioned middle-aged people they have become. Stephen Sondheim has described *Follies* as a reflection of the era where "the country is a riot of national guilt, the dream has collapsed, everything has turned to rubble underfoot."[25] Capitalizing on the setting and on the women's background as performers, the show is set in the now rundown Weissmann Theater on the eve of its demolition and uses musical comedy pastiche to point up the distance that the characters, the country, and the musical theatre itself has traveled since the heyday of the late 1910s and 20s. Onstage, this point was reinforced by staging that had ghostly showgirls drifting around the set and by having the younger alter egos of the women performing the routines alongside their older selves. The sense of nostalgia was increased by casting aging stars of the stage and screen (including Yvonne De Carlo) as the ex-chorines looking back on better days.

This disturbing use of recognizable, reassuring American icons was echoed later that same year when Joseph Papp's Public Theater presented the professional New York premiere of Rabe's play *Sticks and Bones*. Rabe evokes an all-American family modeled (and named) after the stars of the popular American television show *Ozzie and Harriet* only to expose the dark underside of the characters when their son returns from Vietnam with horrifying stories of American transgressions. The all-American geniality quickly peels away to reveal the seething racism, misogyny, and blinkered patriotism of characters in deep denial about themselves as individuals and, by implication, as Americans. *Follies* is in many ways very different from the play, but although the tone of the Rabe play is much darker and more violent than the musical there is an interesting similarity in the way that these shows evoke an idealized past in order to highlight the post-Vietnam shift in the American self-image.

In conclusion, while Prince is rightly acclaimed as a descendant of Broadway directors and producers, his most important legacy may be the way in which he broadened the boundaries of the musical in terms of both form and content. His pioneering work in aligning the musical with current events, with contemporary American drama, and with the techniques and aesthetics of European and Russian theatre directors have resulted in a varied and groundbreaking body of work that is explored more fully in chapter 2.

2. From *Cabaret* to *Sweeney Todd*: Musical Drama on Broadway ⌒

arold Prince's role in establishing musical drama on Broadway can be clearly seen in his body of work from *Cabaret* (1966) to *Kiss of the Spiderwoman* (1993). A closer examination of four early shows—*Cabaret*, *Company* (1970), *Evita* (1978), and *Sweeney Todd* (1979)—reveal the way in which Prince anticipated the director-dramaturgs of the following decades through his active participation in the development process, his need for social relevance, and his incorporation of structural formats and staging techniques beyond the traditions of the American musical theatre. Just as much of American drama in the 1970s was heavily indebted to European playwrights such as Brecht, Beckett, and Pinter, so the structure and staging of many Prince musicals were influenced by a number of foreign directors and theatre traditions. Although these connections have been touched on elsewhere—notably in Ilson—the extent of Prince's interest in theatre outside America has not been extensively analyzed and can usefully be examined here to demonstrate the incorporation of foreign theatre traditions into the Broadway musical even before the so-called British Invasion of the 1980s and 90s.

CABARET

Cabaret is based on John Van Druten's play *I Am a Camera*, which was itself based on Christopher Isherwood's semiautobiographical novels *Mr. Norris Changes Trains* (1935) and *Goodbye to Berlin* (1939). The musical is framed as the experiences of American novelist Cliff Bradshaw who arrives in Berlin just as the hedonism of the Weimar years are giving way to the Nazi era. The central focus is on Sally Bowles, a mediocre cabaret performer who works at the seedy Kit Kat Klub; on the Emcee of the club, whose musical numbers mirror and comment on the growing Nazi presence; and on a cast of regular working-class and middle-class characters who fail—or refuse—to see the menace behind the idealistic rhetoric of a new Germany. When Cliff arrives in Berlin, he is caught up in the nightly parties and the decadent lifestyle of Sally and her acquaintances. By the end of the show, however, the party is over and in the final moments he looks back on the fear that has stunted the characters' lives: Sally has aborted their baby and refused Cliff's offer of a new life in America in order to stay on at the cabaret, and their landlady has turned down her chance of happiness with the Jewish shopkeeper she loves for fear of retribution.

Although Prince's directorial debut was with *A Family Affair* and *She Loves Me* (both traditional musical comedies), *Cabaret* was his directorial declaration of intent and it was this show that signaled his departure from his musical theatre predecessors and contemporaries. Where *Oklahoma!* epitomizes the Rodgers and Hammerstein oeuvre, *Cabaret* clearly illustrates Prince's deviation from their model. Although both shows address some serious issues and reflect the times in which they were created, *Cabaret* was based on a central idea that aimed to challenge rather than reassure the audience. Prince was not the librettist of *Cabaret* (the book is by Joe Masteroff) but he was the initiator of the project and, as producer, it was he who hired the rest of the creative team and guided the development process. Lyricist Fred Ebb has since attested that the songs were to a great extent "made to order" and that he did not really appreciate the significance of the show as a metaphor until it was pointed out to him. Equally, composer John Kander recalls the writing process as a collaborative undertaking led by Prince, with the director's concept influencing the song writing and dictating the overall style.[1]

Prince's impact on *Cabaret* is perhaps best measured against what the show nearly became. As Carol Ilson has pointed out, *I Am a Camera* had long been considered a potential star vehicle for a musical comedy star such as Gwen Verdon. This phantom version held no interest for the more socially minded Prince as he explains in his autobiography: "[Sally Bowles] works in a club, see, and she dances and sings, sings all night long, and it's racy. . . . No point."[2] For Prince, *Cabaret* was neither a star vehicle nor a choreographic showcase but a cautionary tale about the poisonous infiltration of hatred and prejudice into everyday life. His contention was that what happened in 1930s Germany could happen—and to some extent was happening—in 1960s America: "*Cabaret* to its creators was a parable of the 1950s told in Berlin, 1924. To us, at least, it was a play about civil rights, the problem of blacks in America, about how it can happen here."[3]

Earlier Broadway treatments of the holocaust and the spread of Nazism such as *The Diary of Anne Frank* (1955) and *The Sound of Music* (1959) are less confrontational. The former uses theatre as a way of finding some optimistic message of hope and ends with Anne's expression of faith in the human race. *The Sound of Music* uses the war as a frame for a spirited love story and a celebration of individual courage and fortitude, and while there are uncomfortable moments as the Nazi invasion grows nearer the threat is never allowed to overwhelm the story. When *Cabaret* opened a few years later, it was still a sensitive topic but rather than looking back and finding a ray of hope in the devastation of the war, Prince saw *Cabaret* as a platform from which to address contemporary social prejudice.

This active sociopolitical engagement is apparent throughout the development of *Cabaret* as documented by Prince's biographers and by composer John Kander and lyricist Fred Ebb in their book *Colored Lights*. Prince consistently resisted the clichés of musical comedy, devising a story and physical staging that would shock and unsettle rather than reassure the audience. Thus he featured not glamorous showgirls but seedy-looking chorus girls of different shapes and sizes and the Kit Kat Klub was based on a sleazy nightclub that he had frequented while stationed as a soldier near Stuttgart.[4] He also steered composer John Kander and lyricist Fred Ebb toward an authentic score written in the tradition of the jazz and vaudeville songs of the German music hall rather than in the idiom of Broadway musical comedy.[5] This connection to German rather than American performance style was further strengthened by casting Lotte Lenya, a veteran of Weill-Brecht shows such as *The Threepenny Opera*, as the pragmatic landlady. Above all, rather than using the musical to showcase a female star, he was instrumental in shifting the dramatic emphasis from Sally Bowles to the Emcee during the development stages. Where Sally Bowles is the central character in Cliff Bradshaw's memories of Berlin (and in the musical comedy that this might have become), the Emcee is the focal point of Prince's cautionary tale about the spread of prejudice, blurring the line between the onstage and offstage audience as he seduces both into laughing at his prejudiced jokes and tempts them to "leave your troubles outside."[6]

While the thematic impetus for *Cabaret* was race relations in 1960s America, the show's structure and staging were a result of Prince's trip to the Taganka Theatre in Moscow in 1965. From its inception in 1964, the Taganka Theatre had been a magnet for the intelligentsia and artistic community under the leadership of Yuri Petrovich Lyubimov. The production that Prince saw was *Ten Days that Shook the World*, a political revue based on John Reed's book and directed by Lyubimov. *Ten Days* told the story of the final days of the Russian Revolution in 1917 in a montage of scenes using a wealth of theatrical styles and techniques from naturalism to farce and shadow play. It is described by historian Alexander Gershkovich as "an organic unification of theatre, music, pantomime, and eccentricity with vivid marketplace theatre."[7] Above all, Gershkovich emphasizes, "Lyubimov's theatre strived to create a sense of involvement between the spectators and the action. Lyubimov's direction consciously brought the dozing spectator out of his subliminal state of passive contemplation, making him a participant in the performance."[8]

By his own account, it was this aspect of the production that appealed most powerfully to Prince. In his autobiography, he recollects the vivid impact of the staging even though he could not understand the spoken text:

> A curtain of light behind which the scenery was changed . . . paintings on the wall spoke, inanimate objects animated, disembodied hands, feet, and faces washed across the stage. There were puppets and projections, front and rear, and the source and colors of light were always a surprise. All of it made possible by the use of black velour drapes instead of painted canvas. . . . Each of these ideas capitalized on the special relationship of live actors and live observers.[9]

Prince has specifically cited this production as "a turning point in my thinking as a director" and certainly the nonnaturalistic staging, the revelatory use of space, and the direct engagement with the audience have pervaded his work as a director ever since.[10] In *Cabaret*, the impact is clearly seen in the role of the Emcee. The idea of the Master of Ceremonies leading the audience through the show was not original in itself—most obviously, Archie Rice in John Osborne's play *The Entertainer* (1956) was a precursor to the Emcee, continuing to perform Edwardian musical hall numbers even as the British Empire crumbled around him. Prince himself has also acknowledged the similarities with a German Emcee that he saw in a small sleazy nightclub while he was stationed in Stuttgart in 1951: "A dwarf MC, hair parted in the middle and lacquered down with brilliantine, his mouth made into a cupid's bow, who wore false eyelashes and sang, goosed, tickled, and pawed four lumpen Valkyres waving diaphanous butterfly wings."[11] There are clear parallels between this description and the physical appearance of both the Emcee and the Kit Kat dancers in the original production of *Cabaret*. The Emcee's dramatic function in *Cabaret* was more directly due to Prince's encounter with Russian Expressionism. Initially, there was to have been an opening montage of Berlin nightlife performed by the Emcee. However, Prince notes that his night at the Taganka Theatre "showed me that there was another way of structuring theatre. . . . When I came back, we took all those numbers and peppered them throughout the show."[12]

The staging of *Cabaret* also drew heavily on the Taganka's emphasis on audience involvement. Lights were used in surprising ways, blinding the audience at key moments, picking out only the legs of a dancing chorus to produce a nightmarish effect, and creating the division between the real world (events and places in the Sally Bowles story) and the "limbo area" where the Emcee performed numbers that represented shifts in the German mind-set.[13] Designer Boris Aronson's famous tilting mirror (placed upstage to reflect both the cabaret numbers and the Broadway audience watching them) also came out of a desire for audience complicity and involvement in the events onstage. This staging device was reminiscent of both nineteenth-century theatre and of Busby Berkeley's overhead shots in movie musicals, but it functioned here as a menacing reminder to the audience of their own culpability in promoting prejudice. The point was further emphasized by Prince's use of the chorus as an extension of the audience, placing the ensemble on the iron staircase in the "limbo" area of the stage to watch the action unfold as silent collaborators. *Cabaret* was the first of many Prince shows (*Company, Follies, Evita, Sweeney Todd*) in which the chorus was not an anonymous group of singers and dancers but individual characters who served to heighten dramatic tension as narrators and as an extension of the audience. In the following decades, this would become a trademark of many musical dramas, including *Les Misérables.*

COMPANY

As with *Cabaret*, the structure and tone of *Company* can be traced to multiple influences. The thematic echoes with contemporary social drama have already been noted, and Martin Gottfried points out that Prince's dramaturgical abilities played a crucial role in shaping the show: "Few directors could have conceived of a musical made from George Furth's related playlets about marriage. Prince's literary imagination gave him the edge."[14] This is undoubtedly true and as Sondheim points out, *Company* was "the first plotless musical [on Broadway] that wasn't a revue and that ushered in a whole era of shows such as *A Chorus Line*, for example, that depend on vignettes and not on plot."[15] But in addition to Prince's literary sensibility, there are a number of influences in the staging and in the published libretto that can be traced to an interest in foreign artists. Russian-born designer Boris Aronson created a cubist set made up of steel and glass that evoked the feeling of an overcrowded city where people struggle to connect with each other; in his review of the show in *New Republic*, Stanley Kauffman notes

Aronson's debt to the Russian constructionist sculptor, architect, and stage designer Vladimir Tatlin and to the influential Russian stage director Vsevolod Meyerhold who is best known for his work with symbolist theatre.[16]

By contrast, the episodic structure of the show and the sudden changes of tone within the piece can be traced to British director Joan Littlewood. A fervent believer in theatre as a populist rather than an elitist enterprise, Littlewood ran the Theatre Royal Stratford East in London's East End on a shoestring budget and worked largely through improvisation. Most famous for her broad satire of World War I, *Oh What a Lovely War*, she echoed Brecht in her emphasis on social criticism, her working-class perspective, and her use of documentary techniques and episodic structure. The startling juxtapositions and sudden changes of mood and tone in *Company* are particularly indebted to Littlewood's 1959 production of Brendan Behan's *The Hostage* at the Theatre Royal. In his autobiography, Prince notes that he especially admired how Littlewood "cued in fragments of songs—the show had many—and how they erupted from rather than grew out of moments. They had the abrasive effect of attacking when you least expected, creating such life . . . a decade later the songs in *Company* were cued out of a conscious debt to *The Hostage*."[17]

Initially a fairly tightly structured political piece (reminiscent of Behan's earlier *The Quare Fellow*), Littlewood turned *The Hostage* into a broad satire of Anglo-Irish relations using lively snatches of song, direct narration, banners, and music to break down the fourth wall and launch a sensory assault on the audience. At the end of Act One, the hapless young soldier caught up in a farcical British-Irish conflict is pushed around while the company dances wildly. In the ensuing silence, he solemnly sings a song based on the meaningless, pseudo-patriotic platitude that "there's no place on earth like the world."[18] The company then join in and proceed to satirize the easy piety of sentimental songs by subverting audience expectations:

> WOMEN: Never throw stones at your mother,
> You'll be sorry for it when she's dead.
> MEN: Never throw stones at your mother,
> Throw bricks at your father instead.

The bathos and startling subversion exemplified by this song is clearly echoed in *Company*. The score contains several numbers that are in the Littlewood tradition of ironic commentary through unexpected juxtapositions. In "The Little Things You Do Together," the singer's witty and acid reflections on marriage are a recurring counterpoint to a domestic scene. In "Not Getting Married" a disembodied voice sings beautifully of wedded bliss only to be interrupted by the frantic agonizing of the nervous bride-to-be. In "Another Hundred People," *Company* echoes Littlewood's techniques for jolting the audience out of complacency. In this number, a girl who we do not know evokes the incessant movement of New York in a series of documentary, third-hand descriptions of people arriving, departing, and navigating the city. The song functions as a commentary, framing three other scenes and exists outside the narrative storylines of the central characters. It is a frenetic song, difficult to follow, interspersed with only a few extended notes that allow the audience to catch up and it echoes Littlewood's emphasis on stimulating and challenging the audience through unexpected juxtaposition of moods.

EVITA

Prince's work on *Evita* eight years later demonstrated a similar emphasis on social politics and staging that created an interactive and at times confrontational environment for the audience—in this case through the emulation of political rallies and the use of documentary theatre techniques and staging based on German Expressionist cinema.

Evita tells the real-life story of Eva Peron, the Argentinean B-movie actress who married Colonel Peron and became the first lady of Argentina, garnering widespread popularity through her loud advocacy of working-class issues even as she and her husband were accused of corruption by their powerful detractors. A controversial figure, Eva had many political enemies but her early death at the age of 33 helped to solidify her iconic image as a martyr of the people, and her funeral created an enormous outpouring of grief. The musical *Evita* is essentially biographical: starting with the grief and near-hysteria that greeted her death, it takes the audience through her life from her poverty-stricken childhood to her escape to Buenos Aires, her mediocre career as an actress, and her ascent up the social ranks through a series of affairs with high-ranking men. The show is narrated by Che, a Che Guevara figure who punctures the myths around Evita with cynical commentary on her motives and methods, pointing out the discrepancy between her saintly image and the realities of her life and of the Peron regime.

Labeled variously a pop or rock opera (or, in the case of New York critic Stanley Kauffman, a "secular oratorio"), *Evita* is in many ways a descendant of the concept albums epitomized by The Beatles' *Sgt Peppers Lonely Hearts Club Band*.[19] It was already a best-selling album when Prince was hired to turn it into a stage musical, building the show from the lyrics on the album rather than from a traditional libretto. He made a significant shift of emphasis from album to show, making cuts, requesting new material, and conceiving the staging so as to sharpen the show's political angle. It was an approach based not only on stage craft but also on painstaking research and on the first day of rehearsals Prince gave a historical talk to set the events in a sociocultural context.[20]

Prince's emphasis on factual research changed the piece drastically from the more thematically lightweight album to the musical drama that it became. In the notes to the original cast recording, composer Andrew Lloyd Webber and lyricist Tim Rice were at pains to present the story as a fairy tale. To them, this was "a story of people whose lives were in politics, but it is not a political story. It is a Cinderella story about the astonishing life of a girl from the most mundane of backgrounds who became the most powerful woman that her country (and indeed Latin America) had ever seen." Prince's interpretation was far more socially oriented. Just as his approach to *Cabaret* had made it less about Sally Bowles than about the insidious spread of prejudice and hatred, so Prince's *Evita* was less about Eva Peron as a Cinderella figure than about the nature of fame and the media's role in creating icons. The idea of multiple images and reflections was demonstrated even more strongly in Prince's original idea of having every actress in the cast assume the role of Eva Peron at some point in the story. This was narrowed down to three actresses (playing the young Eva, Eva the actress and Peron consort, and finally the mythical, iconic Evita) but a dearth of qualified candidates meant that the role was eventually played by a single actress.[21]

Prince's contemporary, sociocultural approach had a particularly strong resonance in late 1970s Britain where it was developed and produced. Theatre critic and Andrew Lloyd Webber biographer Michael Coveney recalls:

> [*Evita*] seemed to catch something in the politically strife-torn air of Britain in the mid-1970s. This was a volatile, dangerous time in British politics, with a weak Labour government standing helplessly by while the stock market wobbled. The show took its energy from a spirit of football hooliganism, violence, private armies and a growing need for someone to take control.[22]

In particular, there was an obvious point of reference for Prince's exploration of the media's relationship with a woman who married into power and whose references to her modest beginnings were balanced by a steely determination and almost missionary sense of purpose. Shortly after the musical opened, Margaret Thatcher moved into the Prime Minister's residence at 10 Downing Street and there was a sense of life imitating art in the grocer's daughter who took elocution lessons, married a millionaire, and became a dominating political leader. The conservative *Daily Telegraph* even ran a Garland cartoon entitled "Thatchita" showing Thatcher with an Eva Peron hairstyle flanked by two of her cabinet ministers dressed as Peron and Che Guevera.[23] The connection was apparently not lost on Thatcher herself. Of particular interest to the new prime minister was the scene on the balcony of the Casa Rosada in which Eva is suddenly thrust into the limelight and quickly comes of age as a forceful and strident political speaker. As Coveney notes, Thatcher "was not renowned as a theatregoer, but she did like *Evita*. Indeed, she returned several times to watch the Casa Rosada scene from the back of the stalls."[24]

The extent of Prince's influence on *Evita* was partly due to the nature of the project: where a traditional libretto would have stage directions and offer a sense of onstage relationships, the album notes offered only very basic information to enable the listener to follow the story. Interviewed shortly before the opening, Prince explained his aims:

> What I had to do was take the 'Top of the Pops' feeling out of the material. I had to get rid of the sentimentality and give it a harder edge. The story takes place in Argentina, not in mid-Atlantic, and it has to look and sound ethnic as hell. What we are doing is a kind of political rally—that's the force of it. Another word for it is documentary theatre.[25]

The original Che—rock star David Essex—concurred: "Hal is trying to put together something very ambitious, very dangerous, a sort of political show. We don't make any decisions about the politics . . . but it has to make statements, it has to have tension between Che and Eva Peron, it has to have the feeling you get with politics, that huge Nuremberg quality."[26]

The "documentary theatre" idea was reinforced through the actors who were variously narrators and characters in the story and by the character of Che who functioned as a direct link with the audience, often addressing them directly as at a political rally. This distancing device did not sit well with all critics. Walter Kerr of the *New York Times* voiced frustration with the fact that so much of the story was narrated rather than enacted—if not by Che then by the aristocrats, politicians, or other social groups: "It is rather like reading endless footnotes from which the text has disappeared

putting the audience into a kind of emotional limbo. . . . We're not partici-
pants, we're recipients of postal cards (and photographs) from all over.
Which is a chilly and left-handed way to write a character musical."[27]

While the accusations of distancing are well founded, they are in a sense
misleading for *Evita* is not so much a "character musical" as a social commentary. Of all the Prince musicals, this is the show most closely affiliated
with Brechtian "alienation." Prince's dramaturgical and staging emphasis
was as much on the role of the crowd (and, by implication, the audience) in
building an icon as it was on the woman herself. This thematic focus was
realized through a large screen upstage that showed images of Eva Peron and
others to reflect, contrast with, and complement the onstage action. Set
against a relatively bare stage—mainly made up of banners and small set
units such as the bedroom—the enormous images seemed to dominate the
space. Equally, Eva's radio broadcasts took place in front of large prominent
microphones, emphasizing the public nature of her address. Prince also
highlighted the role of Che, who serves as a mirror image of the Emcee in
Cabaret: where the former entices the audience in, Che's interjections are
calculated to undermine the image that Eva is trying to project, inviting the
audience to see through the rhetoric to her questionable motives and
methods. In this sense, it could be argued that Che functions as an onstage
alter ego of Prince himself. Finally, Prince deliberately created distance
between the audience and the different social groups that Eva has to
contend with. On the album, the aristocracy and the military are differen-
tiated only by the fact that the former sing in high-pitched voices while the
latter sing in a lower pitch. Onstage, Prince and choreographer Larry Fuller
turned both groups into ridiculous caricatures of their social roles who moved
around the stage in stylized groups—the aristocracy parading with cigarette
holders held high, while the military moved about in geometric formations.

In addition to emphasizing Eva Peron as a public figure, Prince's production undermined the Cinderella story by deepening the relationship
between the Perons, depicting a complex marriage beneath the iconic
images. In particular, it gave more depth to the rather functional figure of
Peron, making *Evita* the story of the Peron regime rather than simply the
story of a peasant girl's climb to fame. As Prince recalls it:

> I turned each song into a scene, describing to Lloyd Webber and Rice what I felt
> the action should be, what the audience should be seeing. I created a script from
> lyrics by always looking for points of conflict such as Evita's single-mindedness vs.
> Peron's cowardice—he would have quit if she hadn't been pushing, and that kind
> of pulling is terrific in the theatre.[28]

In the original production, this resulted in several striking moments that highlighted the differences between Eva and Peron. Eva's single-minded climb up the social ranks was conveyed through the simple but effective image of revolving doors admitting and then ejecting a series of increasingly powerful suitors; Peron's rise to fame was staged metaphorically as a game of musical chairs, which he wins by sheer chance. When Eva throws out Peron's teenage mistress, the original album's sleeve notes tell us simply that "Eva bursts in to throw Peron's 16-year-old mistress out." In Prince's staging, Peron waited sheepishly outside the door while Eva efficiently ended his affair. And in another telling scene, Prince contrasted film footage of Eva wearing herself out on the Rainbow Tour with the downstage spectacle of Peron balancing two schoolgirls lasciviously on his knee.

Even in the climactic scene on the balcony of the Casa Rosada, where President Peron and his wife address the people for the first time, there was a great difference between the album and the show. On the album, the listener's attention is focused exclusively on Eva's gaining confidence as a speaker. In the production, Prince balanced the image of Eva's transformation from blushing novice to fist-shaking speaker with a change in Peron's bearing toward her: initially paternal, we saw him grow increasingly uncomfortable and annoyed until the final moments when he gave her a perfunctory kiss on the forehead and walked off without her, a hint of the uneasy private relationship beneath the public façade.

Prince's dramaturgical work on *Evita*, turning the Cinderella story into a piece of "documentary theatre" was supported by staging that drew partly on the dynamics of a political rally and partly on his admiration for Expressionist cinema. In her chapter on *Evita*, Ilson briefly cites Prince's debt to *Citizen Kane* and Hirsch quotes Prince's own acknowledgment that his staging of *Evita* was to a great extent inspired by movie director Orson Welles and also German movie director Friedrich Wilhelm Murnau.[29] Murnau is commonly grouped with Fritz Lang and G.W. Pabst as one of the key film directors of Weimar Germany, and film scholar Dudley Andrew points out that while *Citizen Kane* has been lauded for its innovative use of the camera, Welles "is not the first director to express the emptiness of appearances and to question the solidity of the world of ideas. . . . As a master of the long take and the tracking camera he engages a cinematic aesthetic ruled by the great names of F.W. Murnau, Kenji Mizoguchi, and Max Ophuls."[30]

Prince himself has acknowledged that Welles's *Citizen Kane* and Murnau's *The Last Laugh* and *Sunrise* all had "a profound impact" on his staging of *Evita* in terms of mood and visual images.[31] Certainly the structure of *Evita* (established prior to Prince's involvement) follows that of *Citizen Kane*, starting with the death of a mythical figure before telling their story chronologically, pointing out their conflicting impulses and the human story behind the public image. In addition, there is a shared aesthetic between Welles and Prince in their fascination with a crossover between the extended, character-based scenes of the theatre and the cinematic techniques that allow the director to guide and manipulate the audience.

This is particularly true of *Evita*. In his critical study *Orson Welles*, Joseph McBride points out that Welles's background in character-driven live theatre had an enormous impact on his filmmaking, in particular through his lingering shots: "Welles tends to prolong the tension among the characters and camera as long as possible, to approximate the intimacy of a theatrical experience."[32] In a reversal of this, Prince makes use of several cinematic techniques in *Evita* like the close-up and pan, directed lighting, and the idea of manipulating the visual angles from which a particular moment is viewed by the audience. A closer comparison reveals clear parallels in the staging of key moments in *Citizen Kane* and *Evita*. In the musical, Eva's parallel journeys as a political figure and as an individual are depicted through the contrast between large rally-like scenes and private moments that reveal the human being behind the media images. In her first major appearance on the balcony of the Casa Rosada, the blown-up images of Eva on the upstage screen and her position on the raised balcony forces the audience to look up at her from the same angle as the film audience watching Kane for his big political rally. There is also an echo between more private moments in both works. In *Citizen Kane*, our first sight of Kane's ex-wife Susan Alexander is in a shoddy dressing room, leaning her torso over the back of a sofa to look at the journalist (and the camera). In *Evita*, Eva's big speech on the balcony is followed by a small scene in a dressing room. In the original production, she removed her costume during Che's verse and then turned from the mirror to sing her verse, leaning over the back of her chair in a pose very similar to Susan's and, like Susan, focusing our attention on what she has to say through her stillness and the intimacy of the position.

On a technical level, the use of lighting in *Evita* also evokes the aesthetics of Murnau and Welles, using lighting in particular to replicate the use of close-ups and long shots that create the mood and sense of size in the movies. Murnau's *The Last Laugh* tells the story of a proud hotel porter who is demoted to bathroom attendant and thereby loses the respect of his neighbors, his family, and himself. In this narrative, the "chorus" is especially effective in the scene where the Porter arrives home in disgrace, with the intensity of the neighbors' reactions depicted through a series of close-ups of individuals (one woman cupping her ear to listen, another cupping her mouth to call out) and groups (people gathering behind the Porter as he passes). Murnau's use of group close-ups was evoked in the opening moments of *Evita*: after the announcement of Eva's death, there was a total blackout and then, as a picture of Eva Peron appeared on the screen upstage, three groups of people were picked out by spotlights, one at a time, as they started to chant "Evita! Evita!" Murnau's influence can also be seen in the famous logo for *Evita*—the profiled face against a halo of light. One of the most haunting shots in *The Last Laugh* is that of the humiliated and shunned Porter in his bathroom, curled against the wall, trying to hide from the circle of light that picks out his head in the darkness. In the logo for *Evita*, this moment is reversed and the circle of light becomes the halo around Eva's head—captured in her idealized iconic state just as the Porter was captured in his moment of naked degradation.

In *Citizen Kane*, Welles effects a strong contrast between intensely personal moments (like the use of a close-up to imply Kane's feeling of isolation and humiliation during the applause for his untalented wife at the opera house) and broadly public moments (such as the famous wide shot of Kane at a political rally, speaking in front of an enormous image of himself). In *Evita*, Welles's close-up on Kane in the opera box was echoed in the spotlights that created the intimacy of Eva and Peron's first exchange, and later the use of searchlights in the political rally scenes are a reminder of the extraordinary scene in *Citizen Kane* where the guard of a deserted library vault is lit by a single powerful spotlight in the darkness—described by James Naremore as being "caught in a beam of 'Nuremberg' light."[33]

SWEENEY TODD

Where Prince's work on *Evita* added political inflections to an existing work, his major dramaturgical contribution to *Sweeney Todd* was at the conceptual stage where his idea was to take a populist piece of melodrama and turn it into a serious musical drama. The musical is based on a gory Victorian urban myth about a murderous barber whose victims were turned into meat pies. The musical opens with an anonymous chorus singing about the legend, after which the show depicts the chain of events when a wrongly accused convict returns to London and seeks revenge on the man who had him sent away. In the process, he strikes up a partnership with pie shop owner, Mrs. Lovett, and together they hatch a plan to kill lonely strangers who come to Sweeney's barber shop and use them as fillings for meat pies. The story takes a tragic turn when Sweeney learns that his wife was raped by the judge who sent him away and that she subsequently went mad and allegedly died, leaving a daughter behind. Sweeney's attempts to rescue his daughter and wreak revenge on the judge are ultimately foiled through a series of misadventures and in the final moments his quest comes to a somber end as he realizes that his latest murder victim is his crazed, almost unrecognizable wife.

While the book of the musical was written by Hugh Wheeler and the music and lyrics by Stephen Sondheim, Prince was responsible for establishing the tone and thematic emphasis. When Sondheim came across the story at the Theatre Royal Stratford East in London, it was Christopher Bond's new version of the nineteenth-century play that retained the melodramatic elements and invited the audience to boo and hiss. Initially envisioned by Sondheim as an intimate psychological thriller, scored throughout to maintain the intensity, the piece was radically altered by Prince's insistence that it have a wider social context and meaning: "It seemed to me to be relentlessly about revenge and I couldn't afford to be interested in revenge. As a director I need to see metaphor, to find some way of justifying the revenge."[34] Rather than creating a small chamber opera, Prince's ideas turned the story into an epic operatic tale of social injustice. It was this angle on the material that established the tone and scale of the libretto, score, and original staging of *Sweeney Todd* and in assuming this dramaturgical role Prince again foreshadowed the directors of the 1980s and 90s who would not only take on dramaturgical roles but actually be credited as coauthors.

Like *Cabaret* and *Evita*, *Sweeney Todd* became less the story of one person than a commentary on a larger social problem, in this case the dehumanizing effect of the industrial revolution on a society in which all the people are victims of their circumstances: Prince points out that by emphasizing the sociopolitical realities of the story's late nineteenth-century social context "we could say that from the day the Industrial Revolution entered our lives, the conveyor belt pulled us further and further from harmony, from humanity, from nature."[35] This idea was emphasized by Eugene Lee's set design in the original production, which consisted of the reassembled shell of an abandoned factory. Most memorably, the inhumanity of the Industrial Age was emphasized in the efficient system by which Sweeney was able to cut a customer's throat in the barber's chair and then pull a lever to send them sliding down into the furnace of the bake house below, emulating the impersonal efficiency of a factory conveyor belt.

Where *Cabaret* was strongly influenced by Lyubimov, and *Company* by Littlewood, *Sweeney Todd* is indebted to two different European theatre traditions. The macabre and scary elements of the piece are largely indebted to the Théâtre du Grand Guignol that operated in Paris from 1897 to 1962 and became synonymous with a particularly gory form of horror theatre through plays such as *The Ultimate Torture* and *Chop-Chop! Or The Guillotine* that traded on explicit violence and frequently had audiences fainting.[36] However, where Grand Guignol is overwhelmingly serious in tone, pulling the audience into the horror, *Sweeney Todd* the musical follows the tradition (established in the nineteenth-century version by Frederick Hazleton) of combining comedy and horror. The musical deliberately draws attention to the act of turning murder into entertainment by setting the appalling action to music that is a pastiche of cheery British music hall songs. Thus "A Little Priest," in which Mrs. Lovett and Sweeney hatch their plan to sell their murder victims as meat pies, is a comic number featuring a catalogue of Sondheim's most outrageous and witty puns and Prince's slapstick staging.

The distinctive tone of *Sweeney Todd* in the musical theatre canon can also be attributed to Prince's Brechtian impulses. This connection was certainly not lost upon the critics and the Broadway reviews draw comparisons between *Sweeney Todd* and Brecht-Weill collaborations *The Threepenny Opera* (an adaptation of John Gay's *The Beggar's Opera*) and *The Rise and Fall of the City of Mahagonny*, both contemporary pieces that underscore themes of social injustice and use epic staging techniques to turn the audience into active spectators. Responding to these similarities, Clive Barnes of the *New York Post* called *Sweeney Todd* a "folk opera" and Richard Eder of the *New York Times* commented that the piece was "in many ways closer to opera than to most musicals; and in particular, and sometimes too much for its own good, to the Brecht-Weill *Threepenny Opera*."[37] Similarly, *Time* magazine concluded that "this musical is a black-comedy opera with helpings of ha'penny Brecht. Its underlying theme, and *épater le bourgeois* tone, is that man exists only to eat or be eaten by his fellows."[38] There is also a clear connection between the parable of *Mahagonny* (in which man's inherent greed and selfishness leads to the downfall of a city) and the wider social themes of *Sweeney Todd*, with their emphasis on the inhumanity of industrialized society. Brecht's essay on epic theatre was written as a note to *Mahagonny* and he discusses in particular the distinction between traditional "culinary" or "hedonistic" opera (which is intended simply to be enjoyed) and his own work, which combines this pleasurable element with the idea that "the use of opera as a means of pleasure must have provocative effects today."[39] The application of this principle to *Sweeney Todd* is clear as is the echo of *Mahagonny*'s nihilistic tone: *Daily News* critic Bill Zakieriasen pointed out the specific similarities between Sweeney's cry that we all deserve to die (followed by a duet about "man devouring man") and the criminals' contention in *Mahagonny* that "this golden city of Mahagonny we create because all is evil . . . and there is nothing a man can depend upon."[40]

   Prince's approach to the musical as social drama, his literary sensibility, his introduction of new staging techniques, and his interest in formal experimentation did not have an immediate impact on how Broadway musicals were perceived and created. However, his legacy can be seen in some of the most interesting musicals to emerge from London and the American nonprofit theatres in the 1980s and 90s. Directors Trevor Nunn, John Caird, Nicholas Hytner, James Lapine, George C. Wolfe, and Tina Landau differ from Prince and from each other in many ways, but as I show in the following chapters they have all continued Prince's legacy of using the musical to challenge the audience, adopting a thematic and text-based approach and drawing on staging traditions other than those of the Golden Age American musicals.

3. Cultural Barricades: Reading the West End Musicals ❧

The idea of a British cultural "invasion" is not new and the rhetoric surrounding the arrival of musicals on Broadway from London's West End draws on a long tradition of framing cultural exchange in competitive terms. While America led the rise of pop and rock music in the 1950s, the arrival of British pop music in the 1960s led to a sense of being beaten on home territory. Arnold Aronson points out that "the once dynamic American rock 'n'roll had become lifeless and sentimental and was overwhelmed in 1964 by the so-called British invasion of the Beatles, the Rolling Stones, and others, as was the still vigorous rhythm and blues of Motown."[1] Nor was this "invasion" limited to popular music. While postwar Britain was being colonized by American culture, American theatre audiences were exposed to British playwrights such as Harold Pinter, John Osborne, Tom Stoppard, Peter Shaffer, Arnold Wesker, Ann Jellicoe, Robert Bolt, and Edward Bond. Despite the growth of world-class American theatre companies and artists, shows with their origins in the West End or with major British companies such as the Royal National Theatre or the Royal Shakespeare Company (RSC) carry an automatic cultural cachet in America. David Savran notes that in relation to Britain "American theatre retains a colonized mentality" and he explains the prominent place of British drama on Broadway in the 1990s as a result of a deferential attitude: "It would seem that in a time of unprecedented confusion in the United States between high and low, British (or Anglo-Irish) drama appeals to producers, theatergoers, and critics alike because it brings a whiff of elite culture, a touch of class, to American theater."[2]

With musicals, however, this deference has traditionally worked in the other direction and imported Broadway musicals have brought excitement, glamor, and innovation to the West End. However, in the 1980s and 90s, as one hit musical after another crossed the Atlantic from London, the supremacy of Broadway was severely challenged and New York critics voiced the ensuing sense of being besieged. In a 1986 article for the *Sunday Times*, Clive Barnes (the British-born theatre critic from the *New York Post*) discussed the new West End musicals not from an artistic standpoint but with regard to the threat they posed to Broadway's reputation as the world capital of musical theatre: "The Broadway moguls' greatest worry is, quite simply, the new British high-tech musical. These are seen here as a threat to the already ailing Broadway musical."[3] In the *Cambridge History of American Theatre*, Laurence Maslon recalls the joyous celebrations in 1983 when *A Chorus Line* became the longest running musical on Broadway and the occasion was marked by a stunning gala performance involving hundreds of ex-cast members. When *Cats* assumed the mantle 15 years later, the reactions were muted at best:

> On 18 June 1997, when *Cats* became the longest running Broadway show of all time, there was little professional enthusiasm for celebrating a landmark that overturned *A Chorus Line*, the quintessential Broadway musical. In marked contrast to that show's extraordinary celebration in 1983, *Cats* merely roped off the street in front of the Winter Garden for some speeches and a small parade.[4]

The prevailing feeling on Broadway is summed up succinctly by Jack Viertel. Viertel, an ex-theatre critic and dramaturg, is currently the creative director of the Jujamcyn Organization (one of the three large Broadway theatre owners and producers) and the artistic director of *Encores!*, which presents meticulously researched staged readings of rarely performed musicals.[5] While offering reasons why he personally did not take to the West End musicals, he acknowledges that the reaction of the New York theatre industry was not entirely objective: "I think that on the street there was a combination of resentment and admiration . . . there was a tremendous sense that these shows were a step backwards from Prince and Sondheim's best work and the fact that they were more popular than any of our shows turned everyone into a sore loser."[6] Viertel did not enjoy most of the London musicals, but he acknowledges frankly that "I will go to my grave not knowing whether that's resentment or taste."

CRITICAL RECEPTIONS

These feelings of resentment may partly explain the unnecessarily flippant and dismissive tone of some of the New York theatre critics in evaluating the West End musicals. The most popular approach was to sideline the artistic merits of the piece by depicting the shows as technical spectacles in which

dramaturgy, music, staging, and performances were a secondary consideration. Guided by the impressive facts and figures surrounding the shows
(record runs, enormous profits, worldwide productions, record-breaking
ticket prices, sophisticated designs) the popular narrative of this period
classifies the London musicals as an aberration from real musicals, giving
them names that foreground technology and clever marketing rather than
the artistic elements that are usually emphasized in discussions of American
musicals.

While it is true that the London musicals made use of technology to
create exciting visceral effects, the characterization of them as mere spectacles
is often misleading and reductive. The New York reviews of *Miss Saigon* are
a case in point. Howard Kissell opens with a dismissive swipe not only at the
show he is reviewing but also the preceding West End musicals: "First
things first. As everybody knows, British musicals are less about music than
they are about scenery. 'Les Miz' was about The Barricades, 'The Phantom
of the Opera' was about The Chandelier, and 'Miss Saigon,' theoretically
about the consequences of America in Vietnam, is really about The
Helicopter."[7] This opinion of *Miss Saigon* as spectacle was not universally
shared: notably, Frank Rich saw beyond the hype—and the helicopter—to
the powerful musical drama. But Kissel's opening salvo was echoed by
several other leading critics. Edwin Wilson of the *New York Observer*
proclaimed "The helicopter has finally touched down"[8] while Clive Barnes
at the *New York Post* announced "the chopper has landed."[9] The *Village
Voice*'s Michael Feingold was more absolute in his condemnation, opening
his review with the portentous statement that "every civilization gets the
theatre it deserves, and we get *Miss Saigon*, which means we can now say
definitively that our civilization is over."[10] Moreover, this attitude is not
restricted to theatre critics: there is a clear tendency among theatre historians and scholars to adopt reductive terminology in their discussions of the
West End musicals. In *The Broadway Musical: Collaboration in Commerce
and Art*, Rosenberg and Harburg refer to *Cats*, *Les Misérables*, *Carrie*,
Phantom of the Opera, *Aspects of Love*, and *Miss Saigon* collectively as "tech-
nologically staggering megaspectacles"[11] and Barry Singer writes of the
"symbiosis of simplistic melodies, stage spectacle and marketing that characterized the musicals promoted during the '80s by Cameron Mackintosh
(often in tandem with Andrew Lloyd Webber's Really Useful Group), like
Cats or the (strategically) imitative *Les Miz*."[12] This depiction is echoed in
John Bush Jones's *Our Musicals, Ourselves*. Throughout his book, Jones
offers instructive and detailed evaluations of musicals from a sociological
and dramaturgical standpoint, making a strong case for regarding the

musical as more than just light entertainment. However, his chapter on the West End musicals is uncharacteristically dismissive. With the one exception of *Les Misérables* (whose dramatic value he acknowledges) he labels the London shows "technomusicals" that rely "upon theatre technology rather than real content" arguing that "writers of musical theatre—bereft of real ideas and perhaps taking their cue from [the] revolution in telecommunications—turned to spectacle to attract audiences."[13] Equally, he classifies *Miss Saigon* as a technomusical because of the moment in the show when a helicopter lands on stage and because "virtually everyone I spoke to who has seen the show says they mostly went to see the helicopter."[14] Admittedly this technical feat was a major feature of the preshow publicity, but it seems to me that these kinds of criteria (applied by most historians of this era) are problematic in that we traditionally classify musicals according to the qualities of their libretto and score rather than a single moment in the original staging. As *Miss Saigon* is given different productions, should we still call it a technomusical because the original production featured a helicopter? Or will it become a different kind of musical each time it is given a new physical incarnation?

Evita, one of the first British musicals to transfer to Broadway, is an interesting lesson in the questions of ownership and expectations that surrounded the West End musicals in the 1980s and 90s. When the show opened, critics on both sides of the Atlantic claimed it as their own. In London, Sheridan Morley and Milton Shulman welcomed it as a long-overdue homegrown British musical that could proudly compete with Broadway; to Clive Barnes at the *New York Post* the Broadway incarnation was "a definite marker point in the ongoing story of the Broadway musical."[15] This dual sense of ownership was not an accident, for the producers went out of their way to make the musical seem local on both sides of the Atlantic. In Britain, Prince's desire for an experienced American star was overruled in favor of a "search for a star" campaign, mirroring the Cinderella aspects of the Eva Peron story by catapulting relatively unknown British actress Elaine Paige to stardom. For the New York run, Prince realized that in order for this import to succeed on Broadway, New York critics and audiences "had to decide proprietarily that this production was theirs and that they liked it better than they had in London. . . . It's why we went to California and played for four months out of town. It wasn't just this London show, it was now this California show that was making its way to New York."[16]

This decision had a tangible effect on the production itself. In addition to casting Broadway stars Patti LuPone (Eva) and Mandy Patinkin (Che),

he redirected scenes for an American audience. In particular, he strengthened the role of Che:

> Che in London was much more genteel, much more laid back and charming. I
> would say that he was not the driving force in the English production. He was
> really charming, but he was more the observer than the driver. When it came to
> be done in this country, I determined that it should be more political here and
> that Che would have to be the force that drove the story along . . . that American
> audiences would not respond to it—not to the same degree that the
> British had.[17]

THE PROBLEM OF CLASSIFICATION

Subsequent transfers did not tailor their productions in this way, requiring
Broadway audiences to accept them for what they were—imports from the
West End. Clive Barnes's review of *Les Misérables* raises an interesting point
in this regard. After describing the show variously as "magnificent, red-
blooded, two-fisted theatre," "superbly served, instantly disposable trash,"
and a "technical miracle" he advises the reader: "Go with the proper
expectations, and you will have a lovely evening. And before we get stuffy
about it, remember in the good old days you never went to Richard Rodgers
expecting Mozart."[18] Barnes's review implicitly acknowledges the key
problem that faced the new musicals with traditional musical theatre
audiences: the tendency to compare them to something that they were not
trying to be. The ingrained expectations of the Broadway audience were in
large part due to the fact that, since the early days of the musical in the
1920s, when the term was applied to almost anything that contained
singing and dancing, the American musical had gained recognizable thematic, structural, and musical features. Where operetta had been set in
exotic foreign locales, the Broadway musical was almost invariably set in
America and featured quintessentially American characters, themes, and
preoccupations. Structurally and musically, the Golden Age musicals also
started to formalize particular traditions, aided by the fact that they were
developed by practitioners (producers, directors, writers, choreographers,
and performers) who had worked with their immediate Broadway predecessors, ensuring that there was a sense of continuity and shared aesthetics.
By the late 1960s and 70s, the American musical had evolved to a point
where it could be classified and its components dissected in volumes such as
Lehman Engel's *The American Musical Theatre* (1967) and Aaron Frankel's
Writing the Broadway Musical (1977), both of which discuss the

dramaturgy, characters, and song types of the musical based on Broadway musical comedies, musical plays, and concept musicals.

But just as the musical plays of Rodgers and Hammerstein had challenged the traditions of musical comedy, so these new West End musicals required a shift in audience perceptions of what constituted a musical. This was in great part due to the mixed backgrounds of the creative teams and casts. Lacking the musical theatre infrastructure of Broadway, London producers assembled artists whose training was in other areas of the performing arts. Thus the creative team for *Cats* included RSC regulars Trevor Nunn (director), John Napier (set designer), and David Hersey (lighting designer); a choreographer (Gillian Lynne) who had been a classical ballerina before she studied Broadway dance in New York; and a cast whose principals included respected classical actors (Brian Blessed and, until she injured herself in rehearsals, Judi Dench), pop singers and dancers (Bonnie Langford and Sarah Brightman), and a classical ballet dancer (Wayne Sleep). Equally, *Les Misérables* was a collaboration between the RSC and a commercial producer, with a score inflected by European pop sounds and a cast drawn from the RSC, commercial London theatre, and (in Patti LuPone) from Broadway. The result was a varied set of shows, from the episodic, environmentally staged *Cats* and the spectacle-driven *Starlight Express* to the epic social drama of *Les Misérables* and the sociopolitical tragedy of *Miss Saigon*.

OPERA OR MUSICAL?

Classifying these musicals using existing terminology is rendered difficult by the confluence of different musical and dramatic elements that they draw upon. Attempts to label them collectively have largely resulted in terms that emphasize their flamboyant design elements ("technomusicals"), their physical scale and box office success ("megamusicals"), or their use of contemporary pop and rock music ("poperas"). More particularly, there has been an attempt to identify them as either musicals or opera, both of which are problematic labels. As music scholar Joseph P. Swain has pointed out,

> The common labels of "opera," "operetta," "musical," and so forth distinguish not so much the substance of what is going on as the sophistication with which it is going on. All refer to music drama, but the composer of opera will aim to project his drama with the most expansive structures in the most complex musical idioms, while the theater composer works with materials that are inherently simpler.[19]

However, given that sophistication is a rather elusive quality to define, it is perhaps not surprising that there have been divergent ideas about how best to categorize the shows. Viertel notes that on Broadway "there was a feeling (especially among admirers of Sondheim and that whole line of descent from Jerome Kern, George Gershwin, Irving Berlin, Richard Rodgers and Cole Porter) that Lloyd Webber wasn't a composer in the same tradition. Whether it was true or not, that was what people felt."[20] There are several possible causes for this feeling, including the way in which music functions as part of the show: many of the London shows were through-scored, which creates a very different kind of sensibility than the traditional Broadway book musical that features musical numbers interspersed with spoken dialogue. Viertel himself classifies the shows as operettas, comparing *Phantom of the Opera* with *The New Moon* and *Les Misérables* with *The Desert Song*. Jazz musician and broadcaster Steve Race, who mounted a written defense of Andrew Lloyd Webber, also felt that operetta was the natural point of comparison for his work: "He is neither Sondheim nor Gershwin, but he is a fine stage composer in the tradition of Lehar, Romberg, Friml and—yes—Ivor Novello."[21]

The problem of classification arises from the fact that many of the London shows demonstrate qualities of both opera and musicals. As with opera, music plays a central dramaturgical role in establishing mood and conveying the story of shows such as *Les Misérables* and *Phantom of the Opera*. However, like the Broadway musical most of these shows employ colloquial language and accessible music. The resistance to this blend of influences can be traced as far back as 1935 and Brooks Atkinson's review of *Porgy and Bess* in the *New York Times*. Ignoring the often banal recitative of traditional operas, Atkinson objected to the combination of operatic structure and colloquial language:

> Turning *Porgy* into an opera has resulted in a deluge of casual remarks that have to be thoughtfully intoned and that amazingly impede the action. Why do composers vex it so? "Sister, you going to the picnic?" "No, I guess not." Now, why in heaven's name must two characters in an opera clear their throats before they can exchange that kind of information?[22]

Where Atkinson resisted operatic tendencies in a musical, many critics have rejected the term "opera" in relation to the West End musicals. Andrew Lloyd Webber himself has frequently referred to his work as opera and cited Puccini as an inspiration. However, this term met with widespread opposition from the arts community as Michael Coveney records in *Cats on a*

Chandelier: The Andrew Lloyd Webber Story.[23] Noting that "cultural ring-fencing is a favourite pastime of the arts panjandrums, as well as the critics," Coveney cites the passionate objections of leading opera figures to the blurring of the line between the two art forms: "I once heard the opera director Peter Sellars take almost hysterical exception to the suggestion that the musicals of Stephen Sondheim aspired sometimes to the condition of opera. 'Stephen Sondheim does not write operas,' bristled the impatient stager, pronouncing 'opera' as 'opperahs' with a drawling Southern twang— 'Mozart writes opperahs!'"[24] This outrage was echoed by several critics. One extreme example was conservative British cultural critic Bernard Levin's response to the idea that *Evita* signaled the return of opera to the masses. In a *Sunday Times* article, he placed the show in the context of "the false popular culture of our age," which he portrayed as a commercial rather than artistically driven phenomenon with class-ridden references to "adulterating the caviar" and "adding just enough information and novelty to make its prospective purchasers think themselves cleverer than they are."[25] Levin's objections to the show's classification as an opera are not so much on musical or dramaturgical grounds as on the grounds that the theatre, audiences, and colloquial language fail to conform to the cultural elitism that he associates with the art form:

> There is a still greater corruption at the heart of this odious artifact, symbolized by the fact that it calls itself an opera, and has been accepted as such by people who have never set foot in an opera-house, merely because the clichés between the songs ("Let's get this show on the road"—"this is crazy defeatist talk"—"why commit political suicide?") are sung instead of spoken, and the score includes, among the appropriate "Slow tango feel" and similar expressions, such markings as "poco a poco diminuendo."

It is beyond the scope of this book to embark on a musicological analysis of the relative qualities of opera and musical theatre. However, it seems to me that there are interesting parallels in the democratization of both art forms in the 1980s and 90s and that the response to the West End musicals was to some extent mirrored by resistance to changes in the opera world. This is perhaps most obviously seen in the audiences. Musical theatre has always been a popular art form but by the 1980s the Broadway musical audience had grown more specialized. This is not to imply that audiences were necessarily highbrow, although the Prince-Sondheim shows certainly pulled the art form more firmly toward the upper middlebrow. But the divergent paths of show music and popular music, coupled with the move to the suburbs and the escalating price of tickets, helped to create a smaller and more specialized audience for musicals who had a great sense of ownership and loyalty in the traditions of the Broadway musical. Opera and musical theatre both have loyal aficionados who relish the culture surrounding the art form as well as the shows themselves, taking pride in their intimate knowledge of particular shows, performers, and productions.

In the 1980s and 90s, the demographics of the Broadway musical theatre audience began to broaden as the populist appeal of the West End musicals coincided with the rise of the jet plane and the ensuing growth in tourism. This process of democratization and the divergent backgrounds of the creative teams behind the London musicals were echoed in the opera world by the increasing numbers of prominent classical musicians and singers who started to break down the barriers of exclusivity surrounding the art form. New recordings of Broadway musicals such as *West Side Story* and *South Pacific* featured international opera stars Kiri Te Kanawa and José Carreras alongside Broadway's Mandy Patinkin and jazz singer Sarah Vaughan.[26] On British television, soprano Leslie Garrett—best known for her work at the English National Opera—became an ambassador for opera as an accessible art form, projecting a down-to-earth personality and sporting a pronounced regional accent on her frequent television appearances. This was also the era in which opera was appropriated by advertisers and where it became associated with the "low culture" sport of soccer. The biggest opera phenomenon of the 1990s was arguably the packaging of three world-class opera stars (Placido Domingo, Luciano Pavarotti, and José Carreras) as "The Three Tenors." Starting in Rome in 1990, the trio marked the end of four consecutive World Cup soccer tournaments with televised gala concerts that included opera favorites as well as musical theatre classics such as "Maria" and "Tonight" from *West Side Story* and "Memory" from *Cats*. Luciano Pavarotti's recording of "Nessun Dorma" became the anthem of the 1990 tournament while the trio's rendition of "La Donna e Mobile" was the theme song in 1994. In 1992, the theme song of the Olympic Games ("Barcelona") was a duet between Spanish opera star Montserrat Caballé and British rock star Freddie Mercury who went on to record an album together.

The reaction of Broadway audiences to the broad appeal of the West End musicals is not dissimilar to that of opera audiences as the art form was turned into mass entertainment. In "'Expecting Rain?' Opera as Popular Culture," John Story offers an interesting exploration of this phenomenon. His description of the reactions of established opera fans to Pavarotti's free outdoor concert in London's Hyde Park in the summer of 1991 echoes the disdain of Broadway regulars for the tourists who kept the West End musicals running. Story attributes the reaction of the opera fans to a fear that open access was cheapening the art form and cites journalist Kate Saunders's deduction that "the explosion of interest in opera in the 1980s worried people who were attracted by its elitist aura. These are the types who threw away their CDs of Turandot and complain when they hear their plumber whistling 'Nessun Dorma.'"[27] Reporting on audiences who booed a daring new production of *Don Giovanni* at Glyndebourne, Saunders concluded that "their real grievance is, I suspect, the increasing democratization of an art form reserved for the rich." While Broadway ticket prices did not quite match those of the opera, there is a clear parallel in the sense of cultural violation experienced by traditional opera and musical theatre audiences.

REDEFINING "THE BRITISH MUSICAL"

The sense of ownership and cultural attachment that surrounds the Broadway musical has proved problematic in responding objectively to the musicals of the 1980s and 90s. So far, theatre histories dealing with this period have attempted to define the West End musicals within the narrative of the Broadway musical, often pointing out divergences from the accepted "norms" in negative terms. However, this Broadway-centric approach does not take into account the thematic and dramaturgical debt of these musicals to the British theatre from which they and many of their creators emerged. A few contemporary critics have already broached this subject, with many London critics drawing specific parallels between *Les Misérables* and the RSC's (then) recent epic adaptation of Dickens's *Nicholas Nickleby*. In New York, Frank Rich's ability to distinguish among the London musicals set him apart from many of his colleagues. After acknowledging the anger surrounding *Miss Saigon*—largely connected to Equity disputes and high ticket prices—he challenged his readers to "take your rage with you to the Broadway Theater, where *Miss Saigon* opened last night, and hold on tight. Then see just how long you can cling to the anger when confronted by the work itself."[28] While dismissing spectacle-driven shows such as *Starlight Express*, he was more appreciative of the musical dramas but often attributed their success less to the European sensibility or to the British staging techniques than to their echoes of Broadway musicals. Thus he traced the staging of *Les Misérables* back to *West Side Story* and *Fiddler on the Roof*, questioning the British roots of the show through references to the earlier incarnation as a French concert spectacle.[29] Equally, his glowing review of *Miss Saigon* was at the same time a vindication of Broadway musicals: "If *Miss Saigon* is the most exciting of the so-called English musicals—and I feel it is, easily—that may be because it is the most American. It freely echoes Broadway classics, and some of its crucial personnel are old Broadway hands: the co lyricist Richard Maltby Jr., the choreographer Bob Avian, the orchestrator William D. Brohn." He goes on to draw a specific line of connection to Broadway predecessors:

> Without two legendary American theatrical impresarios, David Belasco and Harold Prince, there would in fact be no *Miss Saigon*. It was Belasco's turn-of-the-century dramatization of the *Madame Butterfly* story that inspired Puccini's opera, and it was Mr. Prince who, inspired by Brecht and the actor Joel Grey twenty-five years ago, created the demonic, symbolic Emcee of *Cabaret*, a character that is unofficially recycled on this occasion in a role called the Engineer and played by Mr. Pryce.[30]

It is perhaps stretching a point to trace the success of *Miss Saigon* to David Belasco—something akin to crediting the success of *West Side Story* to William Shakespeare's source material for *Romeo and Juliet*. However, Rich's identification of a connection between *Cabaret* and *Miss Saigon* through the characters of the Emcee and the Engineer is a valid one. It is, moreover, symptomatic of a more profound similarity between two musical dramas that combined Broadway showmanship (slickly choreographed production numbers such as "Mein Herr" in *Cabaret* and "The American Dream" in *Miss Saigon* and emotional star turns such as "Cabaret" and "Her or Me") with more serious sensibilities and a desire to address complicated and painful episodes from recent history in a serious and truthful way.

However, while there are clear parallels between the West End musical dramas and Broadway musicals such as *West Side Story* and *Cabaret*, some elements of the West End shows can also be traced back to the experimental British musicals of the 1950s and 60s. In addition to nostalgic musicals (*The Boyfriend*, *Salad Days*) and show business stories (*Roar of the Greasepaint, Smell of the Crowd*), British musicals of this era included more serious shows that reflected recent developments in British drama, drawing on Brecht rather than Rodgers and Hammerstein for inspiration. The use of songs in *The Hostage* has already been discussed; another prominent example is *Expresso Bongo*, a darkly comic satire on the rock 'n' roll generation produced two years before *Bye Bye Birdie* explored the same subject on Broadway in a more innocent, upbeat way. In addition there was the subgenre that critic Robert Cushman has termed "the Soho musical" depicting working-class and backstreet life as exemplified by *Irma La Douce* and *Fings Ain't Wot They Used T'Be*.[31] These shows are particularly interesting in that, like the subsequent musicals of the 1980s, they had strong ties to developments in the serious nonmusical theatre, having being written and directed by artists who saw theatre as a place for political and sociological debate rather than uplifting entertainment. The British version of the French-originated *Irma La Douce* was directed by Peter Brook, later known for his experimental theatre work. Littlewood followed up *The Hostage* (1958) with *Fings Ain't What They Used T'Be* (1959) and later the bitingly satirical *Oh, What a Lovely War!* (1963)—all of which came out of the same philosophy and approach as her work on sociopolitical plays such as *A Taste of Honey*. Playwright John Osborne made use of music hall pastiche in *The Entertainer* (1958) where the erosion of the British music hall tradition became a metaphor for the changing values of a post-empire Britain; he later attempted a vitriolic (and short-lived) musical, *The World of Paul Slickey*, depicting the life of a critic with all the viciousness of his earlier play, *Look Back in Anger*.

DIRECTORS AND THE NEW BRITISH MUSICAL

In 1983, *The Guardian* critic Michael Billington noted the

> increasing fascination of our top directors with the possibilities of the form. . . . Peter Hall, Trevor Nunn, Terry Hands, John Caird and Peter James all have musicals running in London at the moment and it seems as if a whole generation of post-Leavisite textual puritans are keen to show that they can command the complex logistics of a musical and at the same time say something through the form.[32]

These directors all brought to the musical theatre the techniques and approaches learned from directing in Britain's leading theatres—the kind of classically based training at state-supported institutional theatres that was largely nonexistent in the United States. While the previous generation of British theatre directors had been shaped by the rather austere ethos of the 1950s and 60s epitomized by the Royal Court (emphasizing spare sets and muscular language in plays by John Arden, John Osborne, Arnold Wesker, and Edward Bond) the new generation came of age with playwrights such as David Hare (*Plenty*) and David Edgar (*Destiny*, *Maydays*) whose sweeping State of the Nation plays (tackling broad social issues with large casts, multiple locations, and rapid scene changes) called for fluid and more epic staging techniques. In addition, director David Leveaux recalls the sense of expanding horizons from the increased access to European theatre:

> Suddenly there were these iconoclastic moments where someone would turn up on a trampoline doing *Yerma* and then they'd go back to their country. We became aware that there were other ways, other aesthetics—that it didn't have to be a choice between being anti-aesthetic and puritan about text or aesthetic and therefore careless about text. That it was possible to do both . . . I think there was a conscious effort at that time to deliberately put more spectacular flair into shows. There was suddenly a kind of *visceral* excitement at what you were going to see. This is not to say that we weren't rigorous—there was absolute rigor and attention to text and seriousness about the responsibility to writers. But there was something theatrical going on after a period that you could call anti-theatrical . . . [33]

It was to a great extent this development, rather than preceding Broadway musicals, that shaped the directors of the London musicals in the 1980s and 90s, allowing them to bring new skills and a fresh perspective to the genre. Clearly, some directors and shows were more successful than others. *Jean Seberg* (Peter Hall, 1983) and *Carrie* (Terry Hands, 1988) proved that directorial expertise and the resources of the National Theatre and RSC respectively were not guarantees of successful shows. But at both a practical and symbolic level, the involvement of top British theatre directors with new musicals marked a profound shift, making the case for musicals as an art form that deserved the same level of serious consideration and intellectual engagement as the other performing arts.

So far, the critical debate over the West End musicals has largely been framed as a series of opposing forces: British versus American, opera versus musicals. However, rather than focusing on the deviation from genre norms or framing the shows as cultural missiles aimed at the heart of Broadway, I would argue that we might more productively examine the West End musicals on their own terms and separate the more spectacular shows such as *Starlight Express* from the musical dramas such as *Les Misérables*, *Phantom of the Opera*, and *Miss Saigon*. In analyzing and understanding the latter musicals, the question is not whether the writers know how to emulate Rodgers and Hammerstein or Kander and Ebb; nor is it whether the directors conformed to the staging traditions established by Robbins or Bennett. In order to appreciate the cultural significance of these musical dramas we cannot simply measure them against the conventions of the Broadway musical or opera. Rather, we must recognize their debt to multiple influences—including the British theatre from which they and most of their creators emerged.

4. Beyond the Logos: West End Musical Drama ✄

O ne way of looking at the musical in the 1980s and 90s is that there was a sense of energy in the London shows that can be likened to Broadway's Golden Age. Richard Maltby Jr. points out that

> great American musicals come in bursts. In the late 1940s when I was growing up, there were all these articles in the *New York Times* about the integrated musicals with Frank Loesser writing "you see, the songs will tell the story." And it seemed there was something like that going on in England—that there was a discovery of a different kind of scale of singing and a different kind of show, the impact on an audience.[1]

Not only did Lloyd-Webber and Schönberg introduce a different, more contemporary sound into musical scores, but a new generation of British directors also emerged whose experience in classical theatre and epic social dramas equipped them to tackle musicals using different staging vocabularies than the great Broadway choreographer-directors.

Five West End musicals from the 1980s and 90s are particularly strong examples of musical drama: the Prince-directed *Evita* and *Phantom of the Opera; Les Misérables; Miss Saigon;* and *Martin Guerre*. In this chapter, I look more closely at the contributions of Trevor Nunn and John Caird to *Les Misérables* and of Nicholas Hytner to *Miss Saigon*, exploring the strong impact of their backgrounds in classical drama and opera on their dramaturgical approach to the material. Where Prince drew on British and European theatre traditions, *Les Misérables* and *Miss Saigon* were created by artists who were immersed in these traditions. Like Prince's musical dramas on Broadway, *Les Misérables* and *Miss Saigon* are built around serious themes, have a strong sociopolitical emphasis, draw on "higher" art forms (in this case classical drama and opera), and make use of more traditional musical theatre elements when it serves their purpose. Like Prince, the British directors functioned as dramaturgs and echoed some of his staging techniques, such as the use of the ensemble as an integral part of the story-telling.

LES MISÉRABLES

In the debate over high and low art, *Les Misérables* is a particularly interesting hybrid, characterized by Michael Billington as "high-class musical melodrama . . . three-and-a-half hours of fine middlebrow entertainment rather than great art."[2] Set against the social tapestry of nineteenth-century France, and in particular the brief 1832 Paris uprising, the musical is an adaptation of Victor Hugo's novel of the same name. While it is necessarily an abbreviated version, the story remains largely faithful to the novel. The prelude shows a sequence of three events: a chain gang—mostly made up of petty criminals—hard at work; the escape of one convict, Jean Valjean; and his conversion to Christianity by a priest who protects him from the police even after Valjean has tried to steal from him. The action jumps forward and we see Valjean's journey as a wealthy factory owner and mayor. He is forced to flee his new life when his nemesis—Inspector Javert—hunts him down; subsequently, Valjean lives in seclusion with his adopted daughter, Cosette, whose dying mother, Fantine, entrusted her to his care. On the barricades of the uprising, Valjean spares the life of the captive Javert and manages to save the idealistic young student Marius who loves Cosette. Javert, unable to cope with Valjean's act of mercy, throws himself into the river. In an attempt to protect Cosette from the truth about his past, Valjean leaves her with Marius; however, the lovers track down Valjean on his deathbed, and they are joined onstage by the ghosts of all those who have died before him.

At a producing level, the show was a groundbreaking partnership between a commercial producer, Cameron Macintosh, and the Royal Shakespeare Company (RSC), arguably Britain's most prestigious subsidized theatre company. The partnership was not met with undiluted enthusiasm by cultural hawks, many of whom saw this as a degradation of the RSC's lofty mission. In a 1986 article for *The Guardian*, Nunn noted that *Les Misérables* the musical was greeted "with paroxysms of vituperation from people who saw in it clear evidence that the lights were going out all over Europe and a new dark age had begun."[3] In particular, Nunn confronted his critics by questioning their underlying logic:

> Is there a fear in every generation that if the great mass of people find something pleasurable then for that reason alone, it is unlikely to be of value? . . . I don't

deny the right of any sort of specialist group or theatre audience to exist but I don't myself want to be part of anything that leads the theatre further towards being a minority activity (or an activity for minorities).[4]

From its subject matter to its musical influences, storytelling techniques and the staging of the initial production, *Les Misérables* was fundamentally a product of European theatre traditions. In many ways, it was this—rather than simply the sung-through score or the serious material—that set the show apart and made it seem alien to traditional musical theatre audiences when it reached New York. In his review of the Broadway production, Frank Rich notes different cultural influences, but also traces the show's accomplishments back to Broadway:

> The ensuing fusion of drama, music, character, design, and movement is what links this English adaptation of a French show to the highest traditions of a modern Broadway musical production. . . . In *Les Miserables* [*sic*], Mr. Nunn and Mr. Caird have wedded the sociohistorical bent, unashamed schmaltz, and Jerome Robbins staging techniques in [*West Side Story* and *Fiddler on the Roof*] with the distinctive directorial style they've developed on their own at the Royal Shakespeare Company.[5]

John Caird does not recall a conscious awareness of Broadway templates in the creation of *Les Misérables*, although he acknowledges a possible "kinship" between the show's epic staging and Prince's open staging and fluid storytelling in *Evita*.[6] Instead, he cites a number of European influences, including Ariane Mnouchkine's epic *1789*, the Berliner Ensemble, and the popular theatre of France.

The French origins of the show have been fairly comprehensively discussed in Prece and Everett's essay, "The Megamusical and Beyond: The Creation, Internationalization and Impact of a Genre." They trace its roots to the French Grand Operas of the nineteenth century (exemplified by Giacomo Meyerbeer and Hector Berlioz) that "related some sort of sociopolitical message through a grandiose medium that combined music, drama, dance, lavish costume and set designs, and special effects" and which, like all three Boublil-Schönberg shows (*Les Misérables*, *Miss Saigon*, *Martin Guerre*), were often set against the background of war. Musically, Prece and Everett see the influence of French traditions in all the shows: "Pentatonicism is especially prominent, creating a sense of populist fervour in *Les Misérables*, orientalism (Vietnam) in *Miss Saigon* and medieval French folk music in *Martin Guerre*."[7] In addition, they trace the presence of at least one song of social injustice in every show to the "Gallic musical theatre tradition of audience edification."[8] Certainly all this describes the first incarnation of *Les Misérables* as a French musical spectacular performed in a huge Paris stadium. This first version consisted of a series of moving tableaux depicting key moments from the novel underscored by sweeping emotive music. It was essentially a spectacle rather than a drama, with gaps of several minutes between the different scenes to allow for changing the scenery.

But while this line of descent is one part of the show's heritage, the structure, tone, staging, and thematic emphasis of the musical drama *Les Misérables* was more indebted to Nunn and Caird's background in classical drama. Like Prince, both directors were university educated and they brought to the work a literary sensibility developed by their work on classical texts at the RSC. As dramaturgs, they went beyond Prince to be actually credited as writers (along with composer Claude-Michel Schönberg and lyricists Alain Boublil and Herbert Kretzmer). And while the original idea for the show and much of the music preceded their involvement, it was Nunn and Caird who reconceived it as a musical drama, structuring the narrative, identifying thematic emphases, and selecting and building characters from the enormous social and historical tapestry of Hugo's novel.[9] The structure, design, and staging of *Les Misérables* the musical, created under the auspices of the RSC, was much more reminiscent of their recent epic adaptation of Dickens's *Nicholas Nickleby* than other British period musicals such as *Oliver!* that came out of the traditions of musical comedy and British musical hall. From the outset, *Les Misérables* was treated as a serious RSC project, avoiding what Cameron Macintosh has later referred to as "that traditional, slightly amateurish British aspect of musical theatre where brilliant classical actors let their hair down."[10] (In this respect it was a far cry from Nunn's 1976 *Comedy of Errors* at the RSC, described by Robert Cushman in the *Birmingham Post* as "a cross between classic farce, musical comedy and circus clowning" whose success, according to *The Observer*, depended on "our awareness that this is the famous RSC letting its hair down."[11]) With *Les Misérables*, Nunn and Caird took the Hugo novel rather than the French concert as their starting point. In an echo of Prince's emphasis on unifying themes, Nunn and Caird mined the novel for thematic undercurrents and for details that had been unnecessary in the Paris concert that was intended for French audiences familiar with all the characters and events. In particular, Caird recalls that they identified in Hugo a conflict between social revolution and moral enlightenment.[12] In this thematic framework, the three characters become more than simply protagonists in a story: the struggle between Javert and Valjean becomes a battle between Old Testament vengeance and New Testament forgiveness with Thenardier offering the agnostic point of view. The identification of this theme had a strong impact on the new show, inspiring three of its most powerful and memorable solos in which three philosophical positions are set out: Javert's unforgiving moral outlook in "Stars"; Valjean's prayer for mercy in "Bring Him Home"; and Thenardier's discordant, atheistic "Dog Eat Dog."

Caird confirms that he and Nunn approached the rehearsals and staging of *Les Misérables* as they would any classical drama, down to staging all the scenes themselves with movement based on the "laws of naturalism" (an extension of how we move in real life) rather than using dance choreography.[13] In particular, *Les Misérables* was very much an extension of their recent work on the groundbreaking, eight-and-a-half-hour adaptation of *Nicholas Nickleby*—a fluid, epic production in which an ensemble of actors evoked multiple characters and locations using a few basic prop and costume changes on John Napier's multipurpose set of catwalks and iron railings. The actor-based ensemble work and the nonnaturalistic, constructed (rather than painted) sets that characterized the Dickens adaptation were echoed in the original staging of *Les Misérables*, which was highly fluid and cinematic, using the central revolve to move seamlessly between different locations and offering suggestions rather than full representations of the different locations and social groups.

Despite the common references to technological spectacle, Napier's set for the original production of *Les Misérables* was remarkably sparse, inspired by narrative need. For many of the central battle scenes, it consisted of a barricade made up of chairs and other objects that (although requiring sophisticated technology) retained the appearance of simplicity. Other locations were similarly created by a few simple structures: a table, a couple of chairs, and some silverware constituted the house of the priest whose charity transforms Valjean's outlook on life. A freestanding iron gate on the central revolving platform created the factory yard where Fantine is fired; later, the gate reappeared to divide the inside and outside of Valjean's property in the scene where Marius woos Cosette. Nunn and Caird relied heavily on lighting (designed by David Hersey) to create moods and locations, transforming the stage and revealing space as needed: thus the subtle appearance of lights behind closed venetian blinds created the sense of a cold cityscape for Eponine's poignant "On My Own." Equally, the suicide of Javert as he throws himself into the river (a scene that might normally have invited an elaborate set) was evoked with visual simplicity by having Javert climb over the rail of a bridge set on the floor of the stage and then jump and roll offstage in swirling lights as the bridge was raised up to create the effect of a fall. And just as *Sweeney Todd* had defied Broadway tradition with genuinely drab and austere costumes, so the costumes for *Les Misérables* were more reminiscent of a Brecht play or a nineteenth-century social drama than other period musicals such as *Oliver!*

In the rehearsal room, Caird notes that he and Nunn approached rehearsals for *Les Misérables* as they would any classical RSC show, using theatre games, improvisations, and storytelling techniques inspired by the experimental theatre of the1960s and 70s.[14] Throughout this process, their work was rooted in the novel. In Edward Behr's book on the making of the show, Nunn points out that this was both a historical and a literary project: "On the first day I did a history of all French revolutions, and told them about Victor Hugo's book. The rehearsal period was very exhilarating precisely because it was so unusually serious. There was a lot of improvisational work, and a lot of it came from the novel."[15] Although they did not have the prerehearsal workshop period that they had enjoyed on *Nicholas Nickleby*, they used many of the same collaborative approaches in early rehearsals, trying to give the company a sense of ownership and inviting them to contribute ideas.

Like Prince in musical dramas such as *Cabaret, Sweeney Todd*, and *Evita*, Nunn and Caird made substantial use of the ensemble to create the world of the show and to suggest different social settings. The original cast of *Les Misérables* included actors from a range of performing backgrounds. The Innkeeper was played by RSC regular Roger Allam; Jean Valjean was played by Colm Wilkinson, an Irish singer who had toured Ireland and America since his teens and had sung the role of Che on the original album of *Evita*; and Fantine was played by American singer-actress Patti LuPone who had portrayed Eva Peron in the Broadway run of *Evita*. However, the working method was very much that of the RSC, with an emphasis on improvisation, collaboration, and ensemble work. As Nunn recalls, he and Caird "extracted something like twenty paragraphs from the book which were descriptions of minor characters and we built up a series of individual improvisations, with people switching from one character to another, partly to get people to understand the nimbleness that would be required of them as actors, to populate the scenes."[16]

The original production of *Les Misérables* reflected this improvisational, ensemble approach. The ballad structure of *Nicholas Nickleby*, with actors speaking about their characters in the third person, was echoed in the overhead screens of *Les Misérables* which periodically announced location and year. The rapid character changes in rehearsals also resulted in a strong ensemble in which not only minor parts but also leading roles were doubled up: thus the actress who played the adult Cosette also appeared as a prostitute early on, and the actor playing the innkeeper Thenardier doubled as one of the prisoners in the opening scene. Just as the groups of soldiers, aristocracy, and ordinary citizens created the social world of *Evita*, this emphasis on ensemble playing created a very different aesthetic and visual style than more star-driven musical comedies.

A sample of the show's key scenes reveals the results of this ensemble-based, sociological directorial approach. In the opening moments of *Les Misérables*, a projection announced "1815 Toulon." This gave way to a cloud of smoke at the back of the bare stage from which emerged a line of prisoners in chains. The opening number (the harsh, repetitive "Look Down") established the context of prison labor and the petty crimes that had brought the men there: the stage picture consisted simply of men in rags and chains who stood and knelt in two rows using ritualized, mimed movements to convey the sense of tedium and indignity (see figure 4.1). In a later scene, a few women with washboards suggested the tedium and slog of everyday working-class life. Fantine's descent into prostitution was depicted very simply and symbolically by her final assimilation into the group of whores that she had previously walked past several times. And after

Figure 4.1   Opening scene from *Les Misérables*

the fall of the barricades, the widowed wives and mothers were shown pulling their carts on the revolve—a scene that would not have looked out of place in a production of *Mother Courage*. It was this ability to apply the dramaturgical, rehearsal, and staging techniques of the RSC work to a French musical spectacular that resulted in a show which operates, in the words of Sheridan Morley, "somewhere at the boundaries of not only Hugo but also Dickens and Brecht."[17]

MISS SAIGON

Where *Les Misérables* was a hybrid of different European traditions filtered through the RSC, *Miss Saigon* (written by the same composer and lead lyricist) had a stronger American sensibility. The musical is an adaptation of Puccini's *Madam Butterfly* in which Butterfly, a Geisha, is married to Lieutenant Pinkerton of the U.S. Navy. Pinkerton sails home and she waits for him patiently, refusing to believe that he has abandoned her; when she finally learns the truth, she takes her own life. *Miss Saigon* transposed the action to Vietnam before, during, and a few years after the withdrawal of U.S. troops from Saigon. The lead characters became Kim, a Vietnamese girl forced into prostitution through poverty, and Chris, a U.S. marine who buys her services and falls in love with her. They marry in a local ceremony before Kim's friends, but the fall of Saigon separates them and Kim gets left behind. Several years later, Chris returns to Saigon with his American wife Ellen who sings about his recurring nightmares and the lingering sense of a lost love. Ellen meets Kim and discovers that she and Chris have a child; however, before Ellen can return with Chris, Kim shoots herself to give her son a better life in America—the life that she herself had dreamed about.

Partly as a result of relocating the action, *Miss Saigon* deviates from the opera in significant ways. It is much more obviously indebted to Broadway musicals than to opera, featuring shamelessly extravagant numbers such as "The American Dream" and colloquial American vocabulary for the GIs. In addition, it suggests Vietnamese musical inflections in numbers such as the ad hoc wedding of Chris and Kim. As well as the obvious musical differences, there are some crucial thematic and dramaturgical differences between *Madam Butterfly* and *Miss Saigon*. Co-lyricist Richard Maltby Jr. points out that because the city is locked down, making it impossible for Chris to find Kim once the evacuation begins, it opened up the possibility of a genuine tragic love story—unlike the opera where there is no impediment to the love story except that Pinkerton has no real attachment to

Butterfly.[18] In *Miss Saigon*, Chris does not abandon Kim; they are simply caught on opposite sides of the fence as the Americans withdraw. As a result, there is no obvious villain but rather a tragic story about well-meaning people caught up in the war; the focus shifts from a story about individuals to the wider and more uncomfortable question of the human cost of the Vietnam War.

Miss Saigon was a serious attempt to tackle a pivotal event in U.S. history and the presence of Maltby was crucial in providing an American perspective. He points out that the original script was much closer to *Madam Butterfly*, portraying the American GI as an uncaring villain in the Pinkerton mold. The French writers, he recalls, had a very Eurocentric outlook: "Pinkerton the American was a complete shit. He just didn't care about this girl at all and more than that, none of them [the writers] really understood the impact of Vietnam on the American psyche. It was an attitude of 'we've been losing colonies for years—get over it! What's the big deal?'" By contrast Maltby, as an American, understood the impact of the war on the American psyche in terms of

> this American myth that we don't lose wars, that we always win, that John Wayne would come over the hill and save the day. . . . The city of Saigon was surrounded, we were out and the soldiers that were there still thought that something would happen because it was unthinkable that it wouldn't. And when it suddenly became clear that it wasn't going to happen, it was jaw-droppingly horrendous." His contribution was, in short, "to give them an American sensibility.[19]

Like Nunn and Caird, Nicholas Hytner's background was in subsidized theatre but where the former had mainly worked in classical drama, Hytner also had a background as an opera director. At Kent Opera and the English National Opera, his impatience with the elitism and reverence surrounding opera had resulted in some controversial productions of repertoire favorites and his background in cutting through the clichés to the emotional truth of operas and classical plays had a profound impact on the dramaturgical development and the original staging of *Miss Saigon*. Hytner himself has commented that he sees no great division between the different genres: "The rules of the American musical are not so very different from Aristotle's on how to write a play: where to punch, where to pull back, how to land."[20] His dual background in both drama and opera made him particularly suited to *Miss Saigon*, which retains many of the operatic elements of *Madam Butterfly* including the through-sung score, the heightened emotion, the soaring melodies, and the sheer scale of the action. There is also a different relationship between the singer and the orchestra, as Maltby points out:

> In songs like "Last Night of the World" and "I Still Believe" he [Schönberg] reaches the sung climax and then the orchestra goes on for another 30 seconds. The singers are not singing on the last note of the music. Usually in musicals, the thrill is having the voice cutting off; here, the thrill is the sweep of the orchestra. Luckily he had an opera director who knew how to work with that. Nicholas Hytner had had all the training of European directors—Shakespeare, classics, opera—and he understood music.[21]

Ironically, the moment that came to dominate many reviews and many discussions of the show—the descent of a helicopter during the evacuation of Saigon—had nothing to do with Hytner. The show had already gone through two possible directors (Trevor Nunn and Jerry Zaks) by the time he signed on and the helicopter had been approved by Cameron Macintosh and was under construction as a nonnegotiable set piece.[22] Hytner's chief influence on *Miss Saigon* was to heighten the darker undertones of the piece. As a dramaturg, he functioned much like Prince on *Evita*, identifying problems, making suggestions, highlighting key moments, and ensuring that the themes and the central story were brought out in the script and in the staging. Hytner's approach to staging was textually and thematically based rather than something imposed from outside, and his experience of large casts and ensemble work from opera and plays was invaluable. This was particularly vital given the nature of the libretto. As with *Evita*, the script did not include guidance on staging or even make allowances for the physical necessities of getting sets and people on and off stage. Maltby notes:

> If you read the script of *Miss Saigon*, it reads like a play with four or five big set pieces. I used to joke, what are all those people from the big numbers going to be doing for the rest of the evening? Little did I know that Nick would put the whole city onstage and that even while a scene in a room is happening, people are going by outside and there was this tremendous sense of the life of the city. It was he and [choreographer] Bob Avian who took "The American Dream" and decided to make it this dream sequence which is a climactic moment. Underlying everything in the story is this idea that somewhere in America is the answer, the solution to all of this—so much so that Kim gives her life to make sure that her child has it. So "The American Dream" was both a production number and a very satisfying concluding thought.[23]

As a further illustration of Hytner's ability to manipulate the space and the ensemble, Maltby cites the staging of one of the most emotional songs that might traditionally have been simply delivered downstage as a showstopper: "In the middle of 'Why God, Why?' Chris comes out of the house and is accosted by all the Vietnamese who, as soon as they see an American soldier, say 'get me out of here, get me out of here. Can you help?' And he pushes them away and goes back inside. That addition is completely telling." At no point, Maltby recalls, did Hytner succumb to a need for glamor or shy away from the harshness of the characters' lives. This was exemplified by his restaging of the post-evacuation scene in which the Engineer seeks out Kim: "For 'I Still Believe' someone told Hytner it was odd that this poor girl had a whole house to herself, and he restaged it with twenty people sleeping all over the floor and the Engineer had to find his way through them to Kim. It's a more realistic idea of what her life and the poverty was like."[24]

The physical production, as reported in Behr and Steyn, was based on the same desire for dramatic flow and emotional truth, rather than on creating spectacle for its own sake. The set that designer John Napier had previously developed was changed from the more obvious show business model to something more thematically inspired; Hytner has noted that "it was glossy, glitzy and brutal and I saw the show as more poetic, diaphanous, impressionistic."[25] Tellingly, the staging idea that first defined Hytner's approach was not one of the big production numbers such as "The American Dream" or the march of the enemy troops before an enormous statue of Ho-Chi-Minh, but the bleak nightmare sequence in which the Vietnamese scramble to be air-lifted out before the arrival of the enemy troops, staged with just two walls and a plain wire fence up which the men, women, and children on the outside clambered in a desperate and futile attempt to escape the approaching army of the Vietcong. Hytner later recalled that "I always knew how I was going to do the nightmare sequence and it's the one scene where there's basically just a couple of walls and a fence."[26]

Hytner's emphasis on the more serious themes of the show did not preclude him from using more traditional musical staging. During the development process, he explained that while he was directing most of the piece in a "quasi-naturalistic operatic fashion," he was not restricting himself to this: "The most interesting bits to me are the ones which pull away from that towards more of a 'musical comedy' aesthetic (that's not exactly the right term, but it's the best I can do). I'd quite like to see how much further it can go in that direction."[27] This integration of American musical comedy was, however, tempered by the more European approach to the subject matter and to the piece as a serious drama. This was not a dance-driven show as choreographer Bob Avian soon discovered.[28] The original idea of "Last Night of the World" as a choreographic Fred-and-Ginger number was dropped; instead, it was staged with the two lovers simply holding each other against the background of a city in turmoil, with the small platform they were on disappearing upstage before they had finished singing the song. Just as Prince's *Cabaret* is not ultimately about Sally Bowles and Cliff Bradshaw and just as *Follies* is not really about two unhappy middle-aged couples, so Hytner's dramaturgical suggestions and staging helped to emphasize the tragedy of Kim and Chris as a metaphor for the human cost of the Vietnam War.

The distance between *Miss Saigon* and American musical comedy is perhaps most clearly seen in the show's big production number, "The American Dream." In this song the Engineer gives voice to the Asian ideal of America as a capitalist haven with limitless opportunities where everything and everyone is for sale. There is a great deal of celebratory exuberance in the staging, including glamorous showgirls and an all-American Cadillac car. While the number is indebted to two American artists on the creative team (choreographer Bob Avian and orchestrator Bill Brohn) the tone and the dramatic impact of this number in the show are distinctly un-American. In the song, the usually noble ideals of freedom and equality are transmuted into the Engineer's desire to exercise his "talent for greed." Through his eyes, the American Dream is associated with the nouveau riche ("a new millionaire"), with consumer goods ("Pre-packed, ready-to-wear") and with greed ("Fat, like a chocolate éclair /As you suck out the cream"); in a final anti-idealistic sting, the Engineer points out that "best of all, it's for sale."[29] It is hard to imagine a native Broadway musical giving voice to this view of America. With the possible exceptions of *Pacific Overtures* or *Assassins* (both written by Stephen Sondheim and John Weidman, and the former directed and produced by Harold Prince) the Broadway musical has celebrated America as a land of opportunity for self-starters. In this song, America is more reminiscent of a Brechtian capitalist nightmare, with echoes of the city of Mahagonny, where "all is for sale /And there is nothing that one cannot buy."[30] In Behr and Steyn, there is a very revealing interview with the show's American orchestrator, Bill Brohn. Known for his work on American musicals, Brohn recalls his original setting of "The American Dream": "I had it totally wrong. . . . They'd given me plenty of discussion about the irony in the song, but I'd done it the way I, as an American, would hear a song about the American Dream. Instead, it's a European's view of how an Oriental might think about the American Dream."[31]

In this reading, the Engineer takes on a deeper significance, representing the dark side of capitalism. The character is clearly portrayed as Eurasian and at the beginning of the song, he explains how his cynical and immoral outlook on the world has been shaped by the imperialistic Europeans and Americans who have used his country and his people for their own ends. Tellingly, his reasons for wanting to move to America is the idea that people like him—ambitious, resourceful, immoral crooks—have opportunities to thrive on a much greater scale. He feels that he is wasted as a small-time crook in the East and sees himself as belonging in the more predatory America where "the big sharks feed."[32] In the Engineer, European and American audiences of the show are confronted by an uncomfortable and grotesque reflection of themselves.

It has become the norm to think of the West End musicals of the 1980s and 90s in terms of their scores, impressive sets, and Cameron Macintosh's marketing, all of which have certainly had lasting effects on the way that musicals are written, designed, produced, and experienced. But from an aesthetic standpoint, the contributions of Nunn, Caird, and Hytner in these two musicals have been equally crucial to the development of the musical in this period. They not only continued Hal Prince's pioneering efforts in breaking down barriers between musicals, drama, and opera, but also started to bridge the gap between the different economic sectors, allowing ideas and techniques from their work in subsidized theatre and opera to inform their approach to large commercial musicals. In their staging influences and their unapologetically serious approach to the material, they continued to develop the kind of probing musical drama initiated by Prince on Broadway, repositioning the musical in relation to other performing arts and expanding the parameters of musical theatre as an art form.

5. New Horizons: Nonprofit Musical Drama ✨

J ust as a Broadway-centric outlook has led to the marginalizing of West End musicals in historical narratives, so it has caused musical theatre historians to underemphasize the radical developments in American nonprofit theatres. The 1980s and 90s was the era in which the American musical became a broadly based national art form with regional theatres all over the country starting to develop and produce new works. This shift has had enormous artistic and economic ramifications for the American musical theatre and for the musical drama in particular. The new physical, economic, and aesthetic frameworks of the nonprofit theatres have resulted in an alternative set of constraints and possibilities that has enabled writers and directors to engage differently with audiences and with the art form, using musical theatre to explore complex and sometimes controversial questions about contemporary America. Most discussions of these shows to date have been in terms of the composers and lyricists (or, quite often, the composer-lyricists) as in Prece and Everett's reference to "Adam Guettel's *Floyd Collins*, Jeanine Tesori and Brian Crawley's *Violet* and Jason Robert Brown's *Parade*."[1] The lack of recognition for the directors and librettists as the architects of the new nonprofit musical is misleading for, as with the London musicals, this new kind of show often relied heavily on dramaturgically minded directors. Furthermore, just as writers were often doing double duty as composers and lyricists, so several of the most successful shows had directors who were also the librettists, going beyond conceptual and structural work to actually cowriting the shows. In some cases, the directors were experienced nonprofit playwrights who adopted a creative process that was more collaborative and improvisatory than the usual Broadway production model.

AMERICAN NONPROFIT THEATRE

The gradual shift of new musicals into the nonprofit arena is not an isolated movement: it echoes the earlier shift of new play development away from

Broadway and as such forms part of a larger redefinition of the American theatre. In the 1940s and 50s, just as the British subsidized theatre started to challenge the West End, so the American theatre saw the rise of regional and nonprofit theatres as alternatives to Broadway. In the 1960s, large charitable foundations (Ford, Rockefeller) started to fund nonprofit theatres and in 1961, Ford funding helped to establish the Theatre Communications Group that would provide the theatres with support in areas such as casting, publication, and professional development. In addition, the establishment of the National Endowment for the Arts in the mid-1960s was followed by many state-level Councils on the Arts. The result, as Steven Samuels points out, was "an explosive expansion of art-making and the building of arts institutions, as well as the establishment of arts councils in all 50 states. In a few short years, the not-for-profit arts could fairly be said to encompass America."[2]

Technically, "nonprofit" is an economic term, not an aesthetic one. Specifically, it refers to the fact that an institution or company is funded by federal, state, city, foundation, and individual donors rather than commercial backers; to its tax-exempt status; and to the fact that proceeds from successful shows go back into the theatre rather than into the pockets of the show's investors, as in the commercial producing model. Within this economic framework there is a wide range of theatre: from the inception of nonprofit theatre in the 1950s the term has included a broad spectrum from tiny backstreet companies to multimillion dollar institutions, from the experimental to the mainstream, and from companies that emphasize the actor's craft to those built around writers or directors.

The importance of the nonprofit sector of American theatre can be traced to its divergence from the commercial theatre in relation to aesthetic contexts, audiences, sources of funding, and physical production possibilities, all of which have had an impact on the kind of work that is created. In commercial productions, a particular production team generally comes together for the show and then disperses when it closes. The nonprofit sector is largely made up of companies that have a stated artistic mission and often a stable of associate artists who work together consistently: thus, a given play exists within the larger context of the company's aesthetic, and the artistic leadership will tend to choose work and artists who fit with the overall mission of the theatre. The issue of the audience is also crucial. In the commercial sector, every show is a new start, requiring a marketing campaign to define the show artistically and draw the audience in. The sheer size of the major commercial theatres demands that musicals have a broad audience base if there is to be any chance of recouping costs—a high priority for commercial producers with responsibilities to investors. By contrast, the audiences at nonprofit theatres are often members and subscribers of the theatre. While this can sometimes lead to programming being constrained by a large, expectant subscription base, it also creates audiences who are prepared to try something new as part of their package and who have a loyalty to the institution rather than to individual works.

In recent years, these differences have created some tension as the two sectors of American theatre have become increasingly symbiotic through transfers and coproduction deals. In particular, the nonprofit world has become wary of being defined solely in relation to the commercial theatre, starting with the very name "nonprofit." In "To Have and Have Not," Jaan Whitehead highlights the problem of an artistic sector being defined by

> an economic term with legal implications. . . . The term describes what we are not rather than what we are, what we fail to achieve rather than what we do achieve. . . . What would happen if the situation were reversed, if we were called social institutions and commercial enterprises were called nonsocial institutions? What shifts of perception and power would take place?[3]

However, despite these tensions, it is undeniable that by 1980, the nonprofits had become a vital part of mainstream American theatre, providing an outlet for artists who wanted or needed to work outside the economic and populist constraints of the commercial theatre. They also offered a home to most of the prominent new playwrights including David Rabe, Sam Shepard, David Mamet, Wendy Wasserstein, August Wilson, Marsha Norman, Tina Howe, Emily Mann, Maria Irene Fornés, Lanford Wilson, Terrence McNally, John Guare, and A.R. Gurney. The result, as reported by *Variety*, was that "during the 1963–83 period, Broadway declined from its historic position as the overwhelmingly dominant generator of its own material to the runner-up to the non-profit [and foreign] theatre arena, which now accounts for more Broadway shows than Broadway itself."[4]

THE NONPROFIT MUSICAL

While the nonprofit theatres were helping to breathe new life into American drama, musical theatre writers were still expected to develop their craft in the commercial sector. Jack Viertel notes that "the non-profit theatre had been set up as an antidote to Broadway and Broadway was symbolized by the musical. No-one wanted an antidote to Arthur Miller—they wanted an antidote to Frank Loesser."[5] This was despite the fact that the problems that drove art and artists from Broadway (rising costs, commercial pressures, conservative audiences) also affected musicals. American theatre historian Gerald Bordman points out that by the early 1990s changing demographics and rising ticket costs meant that attending the theatre was "less and less a casual affair. Theatergoers are less willing to risk wasting their money on musicals of dubious merit. The 'modest' hit has become a rarity. Most musicals that do not open to generally laudatory reviews quickly post closing notices . . ."[6]

This is not to say that there were no musicals coming out of the non-profit or experimental world before this. As previously discussed, the 1960s and 70s saw Off-Broadway and Off-Off-Broadway theatres experimenting with new musicals: obvious examples are *Hair* (1967) and *A Chorus Line* (1975), both of which were developed and produced at the Public Theater. *Hair* was in many ways the starting point for the new kind of nonprofit musical, coming out of the same engagement with social politics that led Papp to work with playwrights such as David Rabe and Sam Shepard. Portraying the intersection of the hippie movement with the Vietnam War, and drawing on the sound of the 1960s pop charts, *Hair* gave voice to a social group that had until that point been largely ignored in a musical theatre more interested in looking over its shoulder than in addressing the political and cultural developments of the day. With its episodic and thematic structure, it also managed to break down some preconceptions of what a musical should look and sound like when it transferred to Broadway and London. Interestingly, though, *Hair* was not primarily aimed at the theatre's core audience: on Press Night, Papp sent the cast and crew a memo noting that he was actively seeking a young, energetic audience for that evening rather than the theatre's regular subscribers.[7]

Papp's other great producing success, *A Chorus Line*, is popularly held up as the coming-of-age of the nonprofit musical. Economically speaking, this is true: the creation of the musical through workshops at the Public Theater created a development model that has now become the norm, while the show's lucrative 15-year Broadway run served as a financial incentive to other nonprofit theatres. Artistically, too, it was innovative: the set consisted of a relatively bare stage on which the line of auditioning dancers periodically broke out into musical numbers only to meld smoothly back into the unforgiving line. The structure of the piece and the seeming lack of narrative (the show is largely a series of autobiographical stories told by dancers as part of their audition process) was also fresh if not wholly original, echoing the vignette structure of *Company* five years earlier. But while it was certainly a breakthrough, *A Chorus Line* was not a musical drama in the tradition of Hal Prince: it was at heart a classic show business story, albeit with an ironic reversal in the character of Cassie, a soloist who is trying to get back into the chorus line. Where musicals such as *Cabaret, Company, Follies, Sweeney Todd, Evita, Les Misérables,* and *Miss Saigon* raise complex questions about large social and philosophical issues, *A Chorus Line* essentially celebrates the struggles and exhilaration of being a Broadway chorus dancer.

In the 1980s, however, there was a significant shift. Nonprofit theatres started to develop and produce new musicals as an integral part of their artistic mission, applying the same aesthetic and thematic approach to musicals as to new plays and providing a home for artists and audiences who were interested in more experimental, challenging material. Key to this was the presence of Ira Weitzman at Playwrights Horizons, a New York nonprofit theater organization dedicated to new plays. Weitzman was officially assisting the Artistic Director Andre Bishop but also happened to be keenly interested in the new kind of musicals that had been pioneered by Prince and Sondheim, with *Company* having been a particularly formative experience. But there was little new writing to interest him in the commercial theatre and nothing really happening in the nonprofit theatre where he landed. He recalls:

> It was a pretty abysmal musical theatre scene in the late 1970s. There was hardly anything new being done on Broadway and there seemed to be no avenue for emerging writers. There was barely an outlet for established writers. And so I guess I had the chutzpah of a 20 year old and I went to Andre Bishop and said: "Shouldn't we be doing musicals?"[8]

While there were new constraints (chiefly in terms of budget and technical resources) the writers and directors in the nonprofit theatre were empowered to create musicals whose subjects, storytelling techniques, and themes came from their work on nonprofit drama rather than the Broadway musical. In the 1980s and 90s, major New York nonprofit theatres such as Playwrights Horizons, the Public Theater, and later Manhattan Theatre Club and Lincoln Center, nurtured the kind of work that would have been impossible to create within the economic and aesthetic framework of Broadway. The outcome was a new, experimental body of work that started with *March of the Falsettos* and went on to include *Sunday in the Park With George, Once on this Island, Floyd Collins, The Bubbly Black Girl Sheds her Chameleon Skin, Violet, Bring in Da Noise, Bring in Da Funk, Rent* and *Marie Christine,* many of which can be termed "art musicals" (or, in Wiley

Hausam's term, "anti-musicals") rather than populist works.[9] And crucial to this development was a new kind of director who was able to work within the new artistic and physical parameters of the nonprofit theatres. Key names in the development of nonprofit musicals include Des McAnuff (*Big River*), Graciela Daniele (*Once on This Island, Marie Christine*), and Michael Greif (*Rent*). However, in terms of pioneering the kind of musical drama under discussion here, the most influential figure is arguably writer-director James Lapine.

JAMES LAPINE

Interestingly, Lapine had no sustained interest in musicals, which seemed to him anachronistic holdovers from his parents' generation. Growing up in the 1970s, he recalls that "my generation's music was folk and rock and acid rock so the Broadway music suddenly seemed old to me. It represented what everyone in the counter-culture was desperately trying to get away from."[10] Initially a playwright, his two major plays prior to directing *March of the Falsettos* (*Twelve Dreams* at the Public Theater and *Table Settings* at Playwrights Horizons) were both nonnaturalistic, psychologically based pieces with ties to the theatrical avant-garde. Lapine himself points out that his interest in theatre was with the experimental, downtown crowd and in the more formal investigations of the avant-garde rather than in traditional or conventional theatre.[11] Specifically, he cites his interest in the work of Robert Wilson, Richard Foreman, Meredith Monk, and Richard Schechner. It is telling that, despite clear aesthetic differences, these artists were all concerned with dramaturgical structures in which linear narrative is abandoned in favor of a more fragmented, collage-like text and staging.

This alternative background made Lapine an excellent fit for the new musicals being developed at Playwrights Horizons—a theatre organization whose main stage was designed for intimate plays rather than the big, choreography-driven Broadway musicals and which, therefore, demanded a different approach to storytelling. In *March of the Falsettos* (1981) and *Sunday in the Park with George* (1983), Lapine helped to pioneer nonprofit chamber musicals. These shows, driven by a spirit of exploration and experimentation rather than a desire to entertain a mass audience, offered reflective and intellectually challenging material that deviated from the more brassy commercial musical theatre. Like Prince, Lapine was concerned with using theatre to explore and question rather than simply to entertain and like Prince he found new ways of bridging the gap between the musical and contemporary drama. In addition, the dual function of Lapine as a writer-director on these musicals meant that his aesthetic approach became an integral part of the shows, with each set of skills informing the other in the conception, shaping, and staging of these musical dramas. This was not unprecedented in the musical theatre but it was a departure from the Broadway musicals of the Golden Age in which the two tasks were traditionally performed by different people.

MARCH OF THE FALSETTOS
AND *FALSETTOLAND*

While *Hair* helped to challenge the idea of how musicals should look and sound and highlighted the financial benefits of commercial transfers, the birth of the nonprofit musical drama was arguably the 1981 premiere of *March of the Falsettos* at Playwrights Horizons. Produced in the same year that *Cats* arrived on Broadway, *March of the Falsettos* was a quirky musical drama that combined simple, inventive staging with topical subject matter (homosexuality and the breakdown of the traditional family unit), idiosyncratic characters, witty and subversive lyrics, and a book that made no attempt to pander to popular tastes. *March of the Falsettos* is part of a trilogy that tells the story of a Jewish family.[12] The central character is Marvin who has left his wife (Trina) for a male lover (Whizzer) and is trying to balance his relationship with his new lover, his ex-wife, his son Jason, and his therapist Mendel—who, in an outrageous breach of ethics, falls in love with Trina during counseling and ends up marrying her. In *Falsettoland*, this extended family is completed by the lesbian couple next door and the action turns darker as Whizzer contracts and finally dies from AIDS.

Historically speaking, *March of the Falsettos* was an important milestone in that it prompted a change of artistic policy that would be echoed in theatres across the country. The precursor to the show was composer-lyricist William Finn's quirky song cycle *In Trousers*, which was rehearsed and produced on a minimal budget and was presented in a few late night slots. Weitzman recalls how the songs started to permeate the building, opening up the idea of musical theatre playing a more central role in the institution: "It really started to show us the possibilities of doing substantial work in musical theatre, not just premiums for subscribers or showcases and cabaret."[13] While *In Trousers* was a presented as something of a novelty, *March of the Falsettos* signaled the start of a policy change, making its way through the theatre's development process from readings and workshops to a studio production and then down to the main theatre. Weitzman notes that this production was

> a definitive moment for me, for Bill and for the theatre. Playwrights Horizons had just started a membership subscription program: I think they thought that doing musicals was something they could offer subscribers as a perk, like the little gift in the crackerjack box. But Bill Finn wanted more. He wanted to be treated the same way a playwright would be treated—with respect, with rehearsal time and with proper support—not as a little showcase for a songwriter.[14]

Retrospectively, Weitzman sees it as the start of a new era: "I think it was much more significant than just a little off-Broadway theatre making that commitment. I think it was the beginning of a movement, albeit gradual, of the non-profit theatres being able to embrace new musical work the way they had embraced new drama."[15] This artistic shift was eventually written into the Playwrights Horizons mission statement, turning its initial commitment to new playwrights into a pledge "to support the work of American playwrights, composers, lyricists and librettists."

March of the Falsettos is most commonly associated with Finn, who created the leading characters and established the idiosyncratic, affably anarchic tone of the piece.[16] Like Lapine, Finn's influences included sources beyond the Broadway musical. In addition to musical theatre writers such as Stephen Sondheim and Frank Loesser, he was drawn to the poet Frank O'Hara (whom he describes as "insouciant and conversational and totally personal"), to novelist Philip Roth, and to plays by Anton Chekhov, John Guare, and Wendy Wasserstein in which "you never knew whether it was funny or serious."[17] He was also struck by *Bent*, Martin Sherman's play that deals with the persecution of homosexuals under the Nazi regime, which he found "honest and revealing and gay in a way that I'd never seen." More importantly, though, the show was deeply personal. In the early stages of *March of the Falsettos*, Finn was writing from his own experiences and grappling with issues that touched him personally—a departure from the musical theatre tradition which, perhaps partly due to the collaborative creative process, had tended to deal with subjects in a less personal way. John Bush Jones identifies *March of the Falsettos* as a product of the "Me Generation" and Finn recalls that the show was very much a product of his own bewildered stance toward the openly gay culture that he found himself inhabiting in 1970s New York City.[18] This kind of writing was almost unheard of in the musical theatre as Weitzman points out: "Rarely were musical theatre writers inspired in the way that a playwright might be inspired to take from their own experience or their own lives and fashion that into a play. So Bill Finn and the first few shows of his that we did at Playwrights Horizons were significant in taking the musical theatre into that direction."[19]

Given Finn's deep involvement with the stories and characters onstage, he needed a director who could not only stage his work but also engage in the writing process, teasing out ideas and themes and providing a structural backbone. As Weitzman points out, the kind of collaborator that Finn needed to enable his work to come to fruition had to be able to think like a writer:

> In the case of Bill, if Lapine had not been a writer-director but if he was just a stager he'd still be waiting for Bill Finn to deliver material! It was the writing sensibility that pricked Bill Finn to develop the characters, to make sense of the story and to figure out what he was writing about. So if Lapine had not been a writer/director the production would probably have been less coherent than it ended up being.[20]

In Lapine, Finn found an ideal collaborator. Lapine's talent for dramatic structure and his background as a playwright were crucial to the development of *March of the Falsettos*. After joining Finn on the project, Lapine was instrumental in shaping what Weitzman calls the "intriguing mess" of the early drafts (featuring wonderful episodes but no real story) into a cohesive story that still retained the spontaneity of the original. Finn's eclectic influences and personal tone were combined with Lapine's interest in the theatrical avant-garde and in exploring the psychology and sociology of the family dynamic. The result was a startlingly fresh show that was a sharp departure from musicals at the time. As Michael Feingold noted of *Falsettoland*, "it lives in its time as very few recent productions have, to be enjoyed simply as a picture of the emotional confusions our society is currently living through, and not as somebody's well-meaning lecture on them."[21] Rather than finding order through chaos, or offering the sense of resolution that even the more political Broadway musicals often give their audiences, *March of the Falsettos* and *Falsettoland* imply that emotional confusion is a reasonable response to the contemporary world. Just as social playwrights had been doing for years, Lapine and Finn chose to raise more questions than they answered and to reflect the world around them rather than trying to interpret it through a simplistic lens. Certainly, each show ends with a resolution of sorts. At the end of *March of the Falsettos*, Marvin finds a way to connect with his son and at the end of *Falsettoland*, Whizzer's death sparks a brief finale in which the cast of idiosyncratic characters announce their right to exist on their own terms: "We're a teeny tiny band / . . . / This is where we take a stand. / Welcome to Falsettoland."[22] But essentially the characters' lives are no less chaotic at the end than at the beginning. There is certainly no sense of tying up loose ends ready for a satisfying, rousing finale.

Lapine's interest in experimental theatre rather than Broadway musicals had a clear impact on the development process of *March of the Falsettos* and *Falsettoland*. He adopted an informal, improvisatory approach, allowing themes and plotting to emerge gradually throughout the rehearsals. Rather than working from a completed script or from a directorial concept, Lapine and Finn wrote the piece as they worked with the cast and Lapine notes that "mostly *March of the Falsettos* came out of a certain spirit of discovery and adventure. It was special because it wasn't plotted out. The set came together day by day . . . every day it came in focus a little bit more like a photo."[23] Lapine played a vital role in the gradual structuring and plotting of the show and Finn recalls that he was heavily guided by his collaborator's dramaturgical skills: "He said 'these are the songs you have' and he put them on index cards—the title of the songs and what they were about. Then we got a board and he started plotting."[24] Later on in the process, Lapine would ask Finn to go home and write a song for the scene he was staging the following day. In *Falsettoland*, he identified the need for a song for Marvin's ex-wife Trina, providing Finn with some lines as a springboard; the resulting number, "Holding to the Ground," became the central metaphor for the show. Equally, it was Lapine's idea to have a child in *March of the Falsettos* to humanize Finn's sharp, idiosyncratic writing; the character of Marvin's son Jason subsequently became the lynchpin of the piece as the unintentional victim (although a comic and self-aware one) of the adults' messy relationships. Finn points out that "without the kid—who's the one acted upon—there's no show."[25]

At Playwrights Horizons, Lapine had a modest budget and a tiny stage with no wing or fly space, and the informal, improvisatory feel of his staging was partly a response to these limited resources. The result was that his production allowed the story, characters, and songs to take precedence rather than opulent staging or inventive choreography. Matching the mood of the writing—quirky, fast-paced—Lapine's staging was built around a modular set that consisted of a few basic pieces of furniture on rollers (sofa, chairs, and a freestanding door) that could be easily moved and reconfigured to suggest different locations. The result, as captured in the show's publicity shots, was a slightly frantic, informal, improvisatory staging that matched the funny, neurotic characters and colloquial lyrics. The opening moments established the tone as the four male characters jumped onstage wearing white robes and sunglasses and shining handheld flashlights on themselves as they launched lustily into the irreverent and jaunty "Four Jews in a Room Bitching." Throughout the production the simple, makeshift set and staging lent an air of impermanence that echoed the attempts of the central characters to manage their personal relationships and to rethink the social constructs of marriage and family. As Weitzman observes, *March of the Falsettos* "started to define the chamber musical that the non-profit theatre and Playwrights Horizons primarily began to champion."[26]

SUNDAY IN THE PARK WITH GEORGE

Lapine's next musical, *Sunday in the Park with George*, has received substantially more attention from theatre historians due to his collaborator Stephen Sondheim. Inspired by French artist Georges Seurat's pointillist painting "A Sunday Afternoon on the Island of La Grande Jatte," *Sunday in the Park With George* opens with the character of George painting his restive mistress, Dot, on an island in the Seine, amid the intrusive comments and distracting behavior of passers by. As he paints, George turns the real-life characters into serene inhabitants of his painting: thus an unruly group of mocking boys becomes a band of youthful angels. Everything around him—people, animals, trees—is subjected to the artistic laws of design, symmetry, balance, and harmony. George's relationship with Dot deteriorates and she decides to leave him, carrying his unborn child with her to America. In Act Two, the action jumps to 1984 in America where another George, the great-grandson of Dot and George, has been commissioned to create a piece of mechanical performance art to celebrate the one hundredth anniversary of his great-grandfather's painting. Surrounded by commerce and pressures to service his rich benefactors, this George questions the validity of his art until the ghost of Dot appears and inspires him to keep working and exploring.

With *Sunday in the Park with George* now considered a modern classic, it is easy to forget the controversy that surrounded its original development. Where *March of the Falsettos* had helped to persuade theatres that musicals could be a part of their artistic mission, *Sunday* confronted the preconceptions of people who saw the commercial and nonprofit theatres as separate worlds. It was Lapine who was under commission at Playwrights Horizons and his introduction of Broadway's most revered composer-lyricist into a small nonprofit theatre was a new departure. Weitzman recalls the development process as "walking on eggshells": nobody quite knew what the collaboration would produce.[27] In addition, there was clear resistance from major funding bodies who deemed that a subsidized theatre was no place for an established Broadway musical theatre writer. There is of course a great irony in this, given that the experimental nature of Sondheim's work has always been at odds with the economics-driven Broadway. The irony is compounded by the fact that far from commercializing the nonprofit sector, the show turned out to be one of the most esoteric and stylistically challenging works to emerge in the mainstream nonprofit theatre for years, helping to establish the then-radical concept of a noncommercial musical. *Sunday* was exactly the kind of show that needed the audiences of a nonprofit theatre such as Playwrights Horizons and no responsible commercial producer would have dared to mount it. When the show finally moved to Broadway it was only on the wings of ecstatic coverage from the *New York Times* and in London it premiered at the subsidized National Theatre rather than in the West End.

The aesthetic success of *Sunday in the Park with George* is commonly attributed to Sondheim. In a laudatory essay, Frank Rich proclaimed it as radical as *Oklahoma!* was in its day and noted that Sondheim "has built a bridge between the musical and the more daring playwriting of his day."[28] Although focused more on Sondheim, Rich nevertheless acknowledges the importance of Lapine's background as a playwright of associative dramas, which were "more suggestive of the Sam Shepard school of dreamlike playwriting than either the well-made or Brechtian plays that had determined the shape of past Sondheim musicals." In particular, he points out that Lapine's libretto and direction gave the show "the whiff of a sensibility unknown to the Broadway musical" and identifies the show's debt to different theatre traditions: "*Sunday* . . . blurs old definitions—those that separate Broadway and Off-Broadway, show music and serious music, commercial entertainment and art, the theater and the musical theater."

While *Sunday in the Park with George* is undoubtedly one of the finest artistic achievements in the remarkable career of Stephen Sondheim, it seems important to register the extent of James Lapine's work on the show and thus his role in the development of musical drama. As the show's librettist and director, Lapine's contribution started at a conceptual level and extended not only to the subject matter and themes but also to the development process and physical realization. Lapine credits Sondheim with teaching him valuable lessons about the craft of musical theatre writing.[29] In turn, he brought to the collaboration his own approach to the material that emphasized experimentation and process rather than a highly polished product. His impact can be measured partly by the clear demarcation between the Prince and the Lapine collaborations in the Sondheim canon. Where Prince worked thematically, Lapine worked from specific moments, letting the theme or metaphor emerge through the creative process. As a result, the Prince shows tended to tackle big sociopolitical issues while the latter were more introspective and character based in subject matter and execution. In a 2002 interview, Sondheim explained the differences between the two aesthetics, pointing out that while he and Prince were "Broadway babies" Lapine's background was in Off-Broadway theatre: "Off Broadway [*sic*] has a very different sensibility; it's a much looser way of putting on a show—although James is a meticulous writer and plots very carefully. But his approach to a show, even his approach to writing, is not necessarily traditional—for example, the notion of starting at the beginning. Sometimes he starts in the middle."[30] He notes that he found working with James Lapine "startling" not only because of the creative process but also in terms of how the show was produced: "James said, 'Come on, I have friends at Playwrights Horizons, we'll put it on there.' With Hal it was, 'Okay, I think we can get the Majestic Theatre next fall, so we'll do it then.' It gears your mind differently."

Lapine's influence on *Sunday in the Park with George* is evident in both the subject matter and the development process. The subject of the show (the artist's struggle between art and commerce as exemplified by the story of Georges Seurat) has been overwhelmingly linked to Sondheim by critics and historians. But while there are obvious parallels between the composer-lyricist and the lead character in that both attempt to push the boundaries of their art in a commercially driven industry, the emphasis on the visual arts in the show (with the painter's craft translated into the art of physical staging) owes a lot to Lapine who had started out as a graphic designer. It was, tellingly, an image that served as a catalyst for the project and the Seurat painting that they finally settled on was one that Lapine had previously used in another show. As with *March of the Falsettos*, Lapine's background in more experimental theatre also shaped the development process of *Sunday* and helped to create the distinctive staging of the original production. As with the earlier show, there was no overarching concept but rather an attempt to let the themes and the story emerge gradually. Lapine recalls that there was no outline and he simply started writing, bringing in six pages at a time and eventually starting to discuss what the piece as a whole might look like, where songs might go, and what they might be like. From the outset, Lapine was thinking visually and while designer Tony Straiges rightly won acclaim for his design, the distinctive visual style and the interaction between set and characters was written into the script by Lapine. Where Bennett and Fosse tended to emphasize innovative staging rather than writing, Lapine emphasized both, allowing the character-based writing to dictate the visual realization. Perhaps this is why *Sunday* so intricately merges its different components. Seurat's creative process and vision is evoked in the score (which found musical equivalents of the pointillist painting technique) in the name of the heroine (Dot), and in the physical staging which has become an integral part of the published libretto. Live actors interacted with two-dimensional cutouts of trees, animals, and people to create the sense of reality mixed with fantasy while scenery flew in and out to represent the painting taking shape on George's canvas. The result was an evocation of the world as seen through the eyes of an obsessive painter for whom his surroundings are simply inspirations for his art.

The results of Lapine's input are best exemplified by the Act One finale (see figure 5.1). In this number, all the chaotic and unruly elements of George's real and imaginary life (mainly consisting of people who insist on having independent desires and needs) are brought to order. By the use of

Figure 5.1   The Act One finale from *Sunday in the Park with George*

skillful lighting and staging, Lapine and his designers shaped the painted backdrop, the cutout figures, and the live actors into a recreation of Seurat's painting set to a beautiful musical harmony. Where Bennett and Fosse created memorable theatre through fluidity of movement, the hallmark of Lapine's work on *Sunday* was stillness and a controlled, limited movement vocabulary more reminiscent of Robert Wilson than of Broadway musicals.

In the traditional musical theatre vocabulary, movement—and dance in particular—is the medium through which characters express their emotions and desires. In *Sunday*, Seurat's desire (for a perfect work of art) was realized in still images. The final image of the painting was thus not simply a *coup de théâtre* but also an expression of the artist's joy at realizing his fantasy of transforming real people and events into still images. It was a highly theatrical moment born out of an experimental writer-director's ability to function as both throughout the creative process and represented an important landmark in the development of the new American musical theatre.

6. Nonprofit Directors in the 1990s ᴗ

After the pioneering efforts of theatres such as the Public Theater and Playwrights Horizons, the idea of the serious nonprofit musical spread to theatres across America during the 1990s. While these shows met with varying levels of economic and critical success, the very existence of this alternative home for the art form began to redefine the musical, offering an alternative to both the traditional Broadway musical and the new West End shows. As the economics of the commercial theatre became increasingly forbidding, the nonprofit theatres became vital incubators for musical dramas and nurtured a new generation of musical theatre writers including William Finn, Lynn Ahrens, Stephen Flaherty, Adam Guettel, Michael John LaChiusa, Jason Robert Brown, Ricky Ian Gordon, Jeanine Tesori, and Kirsten Childs. They also helped to produce a new generation of directors whose approach to musical theatre was shaped by the aesthetics of nonprofit drama rather than Broadway musicals. A closer examination of two directors (George C. Wolfe and Tina Landau) and three key musicals from this era (*Jelly's Last Jam*, *Bring in Da Noise, Bring in Da Funk*, and *Floyd Collins*) reveals the central role of a new kind of musical theatre director in creating these shows.

GEORGE C. WOLFE

Like James Lapine, George C. Wolfe came out of the more experimental, progressive theatre and started out as a writer at Playwrights Horizons and the Public Theater. Just as Lapine's aesthetic was shaped by more experimental theatre, so it is the edgier sensibility of the nonprofit sector that has informed Wolfe's work as a writer, director, and as the producer of the Public Theater from 1992 to 2005. However, there are two clear differences between the approaches of these directors. First, where Lapine largely ignored the staging traditions of the Broadway musicals, Wolfe chose to build on them. His directorial role models include Jerome Robbins and Michael Bennett, whose fluid staging and use of space he admires enormously. In particular, Wolfe cites his exposure to Robbins's staging of *West Side Story* as a crucial formative moment: "I was completely, totally transformed by it. I think the quintet in 'West Side' probably liberated a certain aesthetic inside of me—seeing that empty stage and five different realities. I remember leaning forward and watching that sequence in a different way than I watched the rest of the play."[1] Second, where Lapine's work has overwhelmingly focused on white characters and culture (with a particular emphasis on Jewish themes) Wolfe has focused on the cultural politics of race. In particular, he has been a pivotal figure in highlighting the rich, complex, and sometimes painful question of cultural identity in contemporary America, both as the producer of the Public Theater where he championed playwrights such as Suzan-Lori Parks and Nilo Cruz and through his own work from his first musical *Paradise* and his breakthrough play, *The Colored Museum*, to *Jelly's Last Jam*, *Bring in Da Noise, Bring in Da Funk*, *The Wild Party*, and *Caroline, or Change*. In all of these the idea of authentic, complex black experience is held up against familiar stereotypes in an attempt to question the idea of cultural identity as a simple, fixed thing. His approach is perhaps best characterized by a term he used in a 1994 interview for *American Theatre*, where he talked about "silhouettes" as the imaginary snapshots of people that make up cultural stereotypes: "Within a racist, sexist society, we view people without power strictly by their silhouettes. You see the silhouette of a little boy from Brooklyn? Click. That's a hood. Don't have to look inside the person, don't have to listen to what they're saying. You view this silhouette over here, this homosexual? Click. Don't have to know what that's about."[2]

This combination of artistic influences has resulted in some remarkable work on plays—most notably Wolfe's direction of Tony Kushner's epic *Angels in America*—and his exploration of these subjects through musical drama has been truly innovative, continuing the tradition of bridging the gap between musical theatre and contemporary drama. In *Broadway Stories*, Marty Bell reveals the developmental process of *Jelly's Last Jam*, highlighting Wolfe's affinity with social drama and his perpetuation of Prince's legacy as an instigator of musical drama. Noting that contemporary musicals do not often evince the "good old-fashioned outrage at the way we live" that has underpinned so much nonmusical drama, Bell goes on to make two exceptions: "The only contemporary director of musicals who has been able to combine outrage with the show-biz know-how that attracts a large enough audience to fill a Broadway-size musical house is Harold Prince. And now into that territory comes writer-director George C. Wolfe, thirty-seven years old, African-American, with only three off-Broadway plays behind him . . ."[3] There are certainly clear echoes of Prince in Wolfe's philosophy that "in theatre, on one level, you gotta seduce. You're tickling with one hand, and then you're stabbing with the other. That duality is very attractive to me."[4] Crucially, both directors use musical theatre to explore uncomfortable questions using film footage, narration, and montages to create musical dramas that seek to stimulate and provoke the audience at the same time as entertaining them.

Of course, Wolfe is not the first dramatist to explore African-American themes in the theatre and in this respect his musicals *Jelly's Last Jam* and *Bring in Da Noise, Bring in Da Funk* (both of which he cowrote and directed) are in the tradition of African-American playwrights such as Lorraine Hansberry, August Wilson, and later Suzan Lori-Parks. Hansberry's *Raisin in the Sun* was an enormous breakthrough when it premiered on Broadway in 1959 with its depiction of a black working family for whom the simple act of buying a home becomes a social and psychological minefield when it turns out that the only affordable housing is in a white neighborhood where they are not welcome. Another obvious point of reference is August Wilson's decade-by-decade series of plays chronicling the twentieth-century African-American experience. Wilson raises the figure of the black man to poetic heights and chronicles the tragedy of regular working people whose dreams, ambitions, and talents are under constant threat of being derailed, crushed, or mutated by their struggle for recognition in a race-driven society. The 1920s segment of Wilson's chronicle, *Ma Rainey's Black Bottom*, highlights the bittersweet legacy of black entertainers, pointing out the injustices of the recording industry, the racism of the producers who were getting rich off African-American talent, and the lack of real power exercised even by some of the most successful artists. The work of Suzan-Lori Parks in the 1990s has also confronted the idea of African-American identity such as *The America Play* (1990), which tells a stylistically fragmented, poetic story of an African-American everyman and evokes the idea of The Great Hole of History as a criticism of the traditional historical narratives of America that are told selectively from a white perspective.

Jelly's Last Jam and *Bring in Da Noise, Bring in Da Funk* also belong to a tradition of black musical theatre and the depiction of African-American people and culture on the musical stage. Most controversially, the early years of the musical in the nineteenth century included minstrel shows and coon shows, which featured racist caricatures of African Americans as lazy, simple, and childlike with stock characters such as "Mr. Tambo" and "Jim

Crow." The performers (whether white or black) would "black up" using black and white makeup to create a distorted, mask-like version of African-American features. Minstrel shows continued to be a popular form of entertainment—both professionally and as college entertainments—into the mid-twentieth century. However, the African-American presence in musicals also took on more varied forms. In *Black Musical Theatre: From Coontown to Dreamgirls*, Alan Woll traces the long history of "black musicals" on Broadway including the revues of the 1920s; "swing" versions of classics such as *The Mikado*; the all-black casts of *Hello, Dolly!*; race-themed musicals by white writers such as Harold Arlen (*St. Louis Woman, House of Flowers*), Kurt Weill (*Lost in the Stars*), and Oscar Hammerstein II (*Carmen Jones*); Langston Hughes's venture into musicals with shows such as *Simply Heavenly*; and integrationist shows such as *No Strings!* In particular, he notes the sudden explosion of black musicals in the 1970s. These included the controversial *Ain't Supposed to Die a Natural Death* and *Don't Play us Cheap!* (both by Melvin van Peebles); adaptations of plays (*Raisin*, based on Lorraine Hansberry's *A Raisin in the Sun*) and of films (*The Wiz*, based on *The Wizard of Oz*); and, most prominently, a stream of tribute shows to black musicians and performers in musicals such as *Bubbling Brown Sugar, Ain't Misbehavin', Eubie!, One Mo' Time*, and *Sophisticated Ladies*.

It is true that the 1970s saw a resurgence of the Black musical on Broadway, but with a few exceptions these shows largely celebrated the music and performances of black artists without really exploring the painful legacy that lay behind the art. In the 1980s and 90s, there were some attempts to evoke a more contemporary sensibility toward African-American history. In 1982, Michael Bennett's *Dreamgirls* featured a striking number, "Cadillac Car," in which the Cadillac becomes a symbol for upward mobility in America—and the related danger of black art and artists losing their authenticity in the attempt to gain broader acceptance. The song proposes a trade-off that is remarkable in its pragmatic view of the racial cultural divide: "If the big white man can make us think we need his Cadillac to make us feel as good as him / We can make him think he needs our music to make him feel as good as us."[5] In 1994, Prince's revised version of *Show Boat* changed the original Act Two opening number, which was originally an evocation of the 1893 Chicago World's Fair featuring racial stereotypes that were no longer considered acceptable. In the new production, choreographer Susan Stroman replaced this with a choreographic retelling of how African-American dances were systematically appropriated and mutated into "white" dances such as the Charleston.[6]

But while some attempts had been made to reclaim African-American history in the musical, Wolfe's innovation was to treat it not as part of a larger story but as the story itself. In doing so, he referenced and commented on the way in which African-American culture had previously been depicted in mainstream theatre. One particular sketch in *The Colored Museum* highlights Wolfe's debt to and also his distance from his predecessors. "The Last Mama on the Couch Play" is an affectionate send-up of the (now) stereotypical figures in *A Raisin in the Sun* and tackles the question of black "silhouettes" not just as stereotypes imposed by white people but also as clichés perpetrated by black artists themselves. In the Wolfe sketch, the narrator introduces the piece portentously as "yet another Mama-on-the-Couch play. A searing domestic drama that tears at the very fabric of racist America."[7] The "yet another" deliberately undermines the overly earnest tone not only of this announcer but also of the more heavy-handed plays that deal with African-American "issues"—as does the reduction of the African-American matriarch figure to a "Mama-on-the-Couch." After establishing this tone, the sketch systematically satirizes the stereotypes in Hansberry's play from the spoiled, self-important daughter (here a drama student at Juilliard who declaims rather than speaking) and the recognizable figures of the overly pragmatic black mama and the young African-American man raging philosophically against social injustice:

MAMA: Son, wipe your feet.
  SON: I wanna dream. I wanna be somebody. I wanna take charge of my life.
MAMA: You can do all that, but first you got to wipe your feet.[8]

Even in these three lines there are clear echoes of *A Raisin in the Sun*'s central characters of Lena, the practical matriarch, and her son Walter Lee, whose sense of frustrated personal destiny makes him oblivious to more practical everyday problems. The character of "Son" also echoes some of August Wilson's poetic and tragic heroes.

The sketch ends with an irreverent and pointed allusion to the history of black musical theatre. Historically, entertainment has been one of the few professions where African Americans could find acceptance within the white mainstream culture and make a decent wage, although it often meant accepting racist treatment (such as not being allowed to stay in the hotels where they were performing) or having to demean themselves and their cultural heritage through packaging black music and dances for white audiences. This bittersweet history forms the basis for the final moments of "The Last Mama-on-the-Couch Play." After the heightened agonizing of the previous pages, the characters suddenly join in a rousing gospel number that identifies a solution to all the young man's problems: clearly, he should have been born into a black revue where he would have "the chance / To smile a lot and sing and dance."[9] The use of the word "chance" is particularly telling here, satirizing the idea that all a frustrated, downtrodden black man needs is the opportunity to get up and entertain people. The dark side of the African-American song-and-dance tradition is further emphasized in the staging of this number as set out in the published stage directions. The number starts out as a traditional production number with "black-Broadwayesque dancing styles."[10] At the end, the dancing becomes manic and "too desperate to please" until the cast freezes to sing, simply and pointedly: "If we want to live / We have got to / Dance . . . and dance . . . And dance." In the final stage direction, the song-and-dance performers assume "zombie-like frozen smiles and faces" while in the background the audience sees images of coon performers.[11]

In short, *The Colored Museum* sets out the preoccupations and themes that Wolfe's later work on musicals would explore more fully. It signaled an interest in reexamining African-American history by acknowledging the deep social and political significance of black music and dance in the evolution of American culture. In particular, this early work signaled Wolfe's gift for combining humor, wit, and entertainment with unflinching explorations of thorny and uncomfortable subject matter. In the 1990s, these preoccupations were developed in his work as the cowriter and director of two landmark musicals that were developed partly or wholly at nonprofit theatres before transferring to Broadway. What is so striking about *Jelly's Last Jam* and *Bring in Da Noise, Bring in Da Funk* is the savvy, serious, and entertaining way that they mix the traditions of black sociopolitical drama with the storytelling techniques of Broadway choreographer-directors.[12] While Hansberry and Wilson's pieces are primarily cautionary tales and serious social dramas, and "black" musicals have traditionally functioned in the musical comedy and revue mold, Wolfe's work is more in the tradition of musical drama as pioneered by Prince, raising difficult and sometimes uncomfortable questions but presenting them in an entertaining rather than lecturing fashion.

JELLY'S LAST JAM

Jelly's Last Jam tells the story of jazz musician Jelly Roll Morton (the self-proclaimed father of jazz) from his beginnings in a Creole family in New

Orleans through his subsequent success that was tempered by racism directed toward him by white people and which he himself directed at African Americans. The story of Morton is a particularly interesting one in the context of Wolfe's outlook on the issue of race in America. Morton was born into the New Orleans Creole population, a race that was darker than the "white" population but that traced its ancestry to Europe and was considered socially superior to African Americans. In the aftermath of the 1803 Louisiana Purchase, the social laws were changed and Creoles were classed as black. However, many Creoles refused to accept this classification and Morton himself went out of his way to emphasize his European ancestry. As he gained fame, he deliberately distanced himself from the black musicians who had taught him his craft, despite the evidence of their influence in his music.

The original impulse for *Jelly's Last Jam* came from producer Margo Lion who, after an early career in politics, worked first in the nonprofit sector and then as a commercial producer. She recalls:

> My goal was to produce a musical that told the story of how jazz came to be: how this extraordinary art form emerged at the turn of the last century in the city of New Orleans. I wanted to find a good story, to see this moment through the lens of a compelling central character. When I read Alan Lomax's interviews with Jelly Roll Morton, the self-proclaimed inventor of jazz, I knew that I had found my man. Here was a larger than life figure whose intriguing biography and tuneful melodies along with his revered place in musical history made him an irresistible subject.[13]

The musical went through a number of writers and directors in the development process, all of whom Lion rejected because they failed to find a fresh theatrical approach to the material that tackled the complexities of Morton. An early version of the show was attempted by pairing playwright August Wilson with Broadway director Jerry Zaks. However, while Lion is a great admirer of Wilson's poetic language, she points out that he was not the best choice for this project:

> As it turned out, he had seen only one musical in his life, *Zorba*; he had no familiarity at all with the form. The book for a musical is less about literature and more about structure and economy of style; there was no room for the gorgeous language and unhurried pace that characterizes August's writing. By mutual agreement we parted ways after a year of trying to tease out a workable first draft.[14]

Her attempts to find an authentic African-American voice were further thwarted by the fact that most black artists had a very ambivalent relationship with the man who had put their rhythms and sounds on the cultural map, but was also a racist who denied the origins of his art. Margo Lion recalls that "the first playwrights I approached to write the musical reacted viscerally to Morton's attacks on black musicians and rejected my proposal as perpetuating a negative stereotype of African Americans."[15]

Wolfe, by contrast, embraced these conflicting impulses. Instead of trying to resolve the conflict, he made it the engine of the show. Lion recalls:

> It was only when I found George Wolfe that I discovered a writer intrigued by the dramatic potential inherent in the conflicts and prejudices that existed inside the black community. His vision was far more expansive and compelling than my original notion of keeping the musical in New Orleans and focusing on the less complicated and more surface entertainment value of Morton's music and his self-aggrandizing and colorful persona.[16]

As he had done in *The Colored Museum*, Wolfe structured *Jelly's Last Jam* as a montage of connected scenes rather than the continuous narrative associated with musical comedy and musical plays. The result was a powerful musical drama that was viscerally thrilling, thematically provocative, and intellectually challenging. Some critics condemned it for not being the kind of linear musical that it was trying to break away from. Most damningly, Rex Reed of the *New York Observer* dismissed it as "a plotless, meandering trifle" and a "dark, confusing jumble," before asking sourly "how many people will be willing to pay $60 for an orchestra seat to watch some tap dancing?"[17] In fact, had this simply been a tap dance show it might have been an easier sell to a mass audience. However, this reaction was not echoed by other leading critics. In the same paper, John Heilpern claimed that "its theme alone . . . propels the Broadway musical into a new, and adult, age" and that "it transcends and demolishes the black stereotypical Broadway revue . . . *Jelly's Last Jam* is not, in its ultimate achievement, a 'black musical.' Rather, it is a great musical—impure and simple."[18] In the *New York Times*, Frank Rich's review famously concluded that, despite its imperfections, this was a show that "anyone who cares about the future of the American musical will want to see and welcome."[19] Rich highlighted the dramatic power of the piece and the serious thematic content framing Wolfe as a "visionary talent" who had given the show "ambitions beyond the imagination of most Broadway musicals."

In an echo of Prince's emphasis on overarching social themes, Wolfe's depiction of Morton became a metaphor for America at large with Morton's life as an allegory of the human cost of racial and cultural denial. The show starts at the end, just after Morton's death where he is greeted by an angel—the "Chimney Man"—who takes him back through his life and makes it clear that unless he acknowledges his cultural roots he will not be allowed into Heaven. The audience is taken on a selective tour not only of Morton's life but of the people and cultural influences around him. In two early scenes we see Young Morton (played originally by Savion Glover) subjected to two very different cultures. In one scene, his Creole ancestors sing down at him in a stilted version of classical European opera, staged as heads in picture frames looking down from the walls and telling him that he is of European descent. Another scene depicts the vibrant weekly New Orleans market where the black community gathered and where the young Morton absorbed the African rhythms and cadences of the vendors. The show is fundamentally constructed as an argument aimed at persuading Morton of his debt to African-American culture and by the end he acknowledges of his onstage alter ego that "inside every note of his is what he came from . . . what he is."[20] He is finally admitted to heaven as part of the musical line that he formerly denied: "Go forth Armstrong! Go Forth Ellington! / Go forth Basie, Bolden 'n' Bechet! / Go forth Morton!"

Jelly's Last Jam had its first incarnation at the nonprofit Mark Taper Forum and was the result of collaboration between commercial and nonprofit theatre. Lion points out that the opportunity to workshop the piece in a nonprofit venue was crucial to the development of an experimental musical drama:

> *Jelly* could only have been developed in a partnership with a not-for-profit theatre. Given that the show was an exploration of a highly complicated and unconventional African-American central character, one who was not easily recognizable to a traditional Broadway musical audience, George needed the opportunity to work through the material in a protected environment and with an audience that was used to seeing new work. Of course, given the fact that this was George's first outing as a commercial director and that he had never directed or written a musical (his other hat), there was little hope of interesting commercial co-producers without an earlier production with good reviews and positive feedback.[21]

The convergence of commercial and nonprofit sensibilities clearly informed the original production, in which Wolfe's serious thematic emphasis was balanced by the celebration of African-American culture—especially through the central performances of tap dance phenomenon Savion Glover as young Jelly and the charismatic Gregory Hines (also an acclaimed tap dancer) in the lead role. The emphasis on star performers in both *Jelly's Last Jam* and *Bring in Da Noise, Bring in Da Funk* is particularly interesting in that it is more reminiscent of musical comedy than musical plays or musical dramas. Musical comedies are often at least partially tailored to their stars and have in some cases been created as a star vehicle, such as *Chicago* (which was created as a vehicle for Gwen Verdon) and most of Ethel Merman's shows. But while the original performers can make a great impression in a musical play or drama it is finally the show and the production itself that is the biggest star. In *Jelly's Last Jam* and *Bring in Da Noise, Bring in Da Funk*, Wolfe managed to pull off a balancing act between the star-turn of musical comedy and the thematic and narrative drive of the musical drama. He achieved this by working the showmanship, charisma, and dance talent of Savion Glover and Gregory Hines into the story itself. *New York Observer* critic John Heilpern noted that in Wolfe's vision "a 'turn' as apparently and mindlessly entertaining as tap, takes on an emotional explosiveness and fever here way beyond showbiz ritual. Tap dance becomes memory play, duet for drums, a danced narrative."[22]

This duality between show business and serious drama, implicit in the libretto of the show, was also evident in Wolfe's staging of the original production, especially in relation to Jelly's first entrance. In a classic musical theatre device, Jelly rose up to the stage on a platform during the opening number, but the entrance was muted by the fact that he was hidden by the chorus. As Frank Rich noted, this entrance foreshadowed the show's darker themes: "Mr. Hines arrives without fanfare. His back is to the audience, his posture crestfallen. When he finally turns to look at us, he is unsmiling, mute, and shuddering. His baggy eyes are wide with the fright of someone who has just seen a ghost."[23] In fact, the ghost Jelly is confronted with is his younger self. When he makes his entrance he is already dead and what we are witnessing is not just a celebration of Morton's jazz but an indictment of the racism that led him to deny its roots in African-American culture. By opening the show with an explanation of Morton's metamorphosis into a destructive, self-loathing racist, Wolfe "tickled" and "stabbed" the audience simultaneously, creating an uncomfortable frame for the otherwise joyful tap dancing number between Morton and his young alter ego based on the African rhythms he has learned at the market. This push-and-pull approach to the audience also underscores the most devastating moment in the show when, at the height of his success, Jelly turns on his best (African-American) friend in a fit of unjustified jealousy and humiliates him by handing him a doorman's jacket. In Wolfe's original production, Jelly's ensuing self-eulogy "Dr. Jazz" was shockingly staged with dancers in blackface. As before, Wolfe balanced the visual pleasure of Hines's showmanship with a clear reminder of the degrading practice of blacking up.

BRING IN DA NOISE, BRING IN DA FUNK

In the production credits for *Bring in Da Noise, Bring in Da Funk*, it states that the book is by Reg E. Gaines, the lyrics by Gaines, George C. Wolfe and Ann Duquesnay, and the music by Daryl Waters, Zane Mark, and Ann Duquesnay. Initially, however, the show was born out of a conversation between Wolfe and Savion Glover and both men were instrumental in its development, with Wolfe being credited with "conceiving" the show as well as directing the original production, which was heavily built around Glover's choreography. In *Bring in Da Noise, Bring in Da Funk*, Wolfe again tackled questions of cultural heritage and racial identity through the vocabulary of black music, going even further in the interaction between written text and staging. Given the subject matter, the hard-hitting social commentary and the formal experimentation in this piece, it was imperative that the show be developed within the artistic parameters of a nonprofit theatre rather than the commercial marketplace, and the show was initially produced at the Public Theatre before transferring to Broadway. *Bring in Da Noise, Bring in Da Funk* is more loosely constructed than *Jelly's Last Jam*, constituting an episodic journey through African-American history using sketches, dance, and voice-overs to depict the cultural politics of African-American music and dance from the days of slavery to the present. While the show celebrates African-American dance—and especially tap dance—it combines showmanship with the social criticism and narrative techniques (visual and aural juxtapositions, projections) of Brecht, Littlewood, and the Sondheim-Prince collaborations. Like his predecessors, Wolfe takes familiar and well-loved traditions (here, tap dance, and popular music) and places them in a new and discomfiting context. In the original production, he used film projections, slides, and thematically linked scenes to offer a kaleidoscope of the African-American experience through the language of music and dance.

The musical is subtitled "A Rap-Tap Discourse on the Staying Power of the Beat" and the original production opened with two simple projections: "In the beginning there was" . . . "da beat!" Throughout the show, "da beat" became a central metaphor for African-American identity. The show opened with the depiction of Africans being brought to America as slaves. Against the projected image of a calm sea with ships, the audience saw a box onstage with a man inside it; as a singer listed the names of slave ships, the man started to tap a rhythm with his foot. Later, in one of the most memorable sections of the show, iconic Hollywood stereotypes of African Americans are exposed against a projection that asks pointedly: "Where's the Beat?" One scene satirizes the scene in Disney's *Dumbo* where black crows on a washing line sing minstrel songs: in *Noise/Funk* this became a series of puppets on a washing line doing watered-down versions of African-American dances. The Shirley Temple—Bill "Bojangles" Robinson partner-ship is parodied in a scene where a white girl named Li'l Dahlin' (a reference to the Neal Hefti classic made famous by Count Basie) perkily asks her black adult colleague: "Uncle Huck-A-Buck, tell me another thing / Why do I make more money than you?"[24] Wolfe staged this using a puppet for the white girl who was operated by an anonymous black performer, point-ing up the humiliation of the Bill Robinson character by the fact that he was being patronized by an inanimate object (see figure 6.1). The black enter-tainers who allowed themselves to be portrayed in this way are also critiqued in a scene that has Uncle Huck-a-Buck singing: "Who de hell cares if I acts de fool / When I takes me a swim in my swimming pool?"[25] Wolfe had the black performers strumming banjos dressed in costumes that evoked minstrelsy—an uncomfortable reminder for the audience and also a vivid illustration of what the lyrics are saying.

The central idea of "da beat" as a metaphor for the African-American spirit was concretized through Wolfe and Glover's original staging. In one scene, the voice of a woman describing the wonderful conditions "up north"

Figure 6.1   Scene from the Hollywood section of *Bring in Da Noise, Bring in Da Funk*

was set against the image of dancers as pieces of machinery in a Chicago factory and the sound of rattling chains and banging on steel. The dancers became cogs in a huge, inhuman machine, bringing the infernal factory to life through feet tapping and drumming. In another scene, resistance to an announcement of a South Carolina law outlawing the use of drums by slaves was conveyed by the sound of a defiant solo singer, one quiet drum, and a tapping of feet. In a third scene two dancers, lacking formal instruments, made music on saucepans that they had attached to themselves, in effect becoming a musical instrument. Toward the end of the show, a section entitled "Street Corner Symphony" presented virtuoso street dancing followed by a section called "Conversation" in which the image of contemporary street dancers was paired with voiceovers of people relating how dance and music helped to give them a sense of cultural identity as both Africans and Americans.

Given the metaphoric significance of the beat and of the titular "noise" one of the most dramatic moments was at the end of the section depicting a lynching. Staged through multiple sensory devices, this section included a visual projection listing the paltry crimes for which African Americans were condemned (including brushing against a girl or slapping a boy) while onstage the lynching was represented by a dancer in a spotlight on a raised platform with his hands behind his back. As the life ebbed out of him, his tapping gradually slowed and we knew that he was dead when it finally stopped. In a world where "da beat" denoted the African-American spirit, silence signified death.

FLOYD COLLINS

By the early 1990s, there were three main models for prolonging the life of a musical beyond its original run: what we might call the "classic touring model" (where the original production is taken out on tour), the "transfer model" (in which a nonprofit show is picked up by commercial producers), and the "cloning model" (where the original production is reproduced in concurrently running productions across the world—an approach perfected by Cameron Macintosh in the 1980s and 90s). For nonprofit theatres, the transfer possibility has become an important financial consideration and for the more commercially viable musicals it has proved a useful source of income to be ploughed back into the theatre's budget. However, this increased dialogue between the nonprofit and commercial theatres has prompted new ethical and aesthetic questions around how far musical projects are chosen with an eye to their transfer potential. It also raises the question of what happens to musicals that are not inherently commercial after the first production. As the nonprofit theatres became increasingly active in musical theatre development, a fourth model of distribution arose that by-passed the commercial theatre completely. Rather than transferring or cloning the iconic original production, some nonprofit musicals were given a series of new productions in nonprofit theatres across the country. This new model of distribution—echoing that of new plays—recognizes the growing number of musicals whose scale, tone, or subject matter is more suited to the aesthetics, spaces, and audiences of the nonprofit theatres than to the large commercial houses.

One early example and beneficiary of this new system is *Floyd Collins* (1993). The show is based on the real-life story of a young Kentucky man who, in the winter of 1925, got trapped in a mine when searching for caves that could be made into tourist attractions. As the rescue operation dragged on, the American media descended on the scene and for two weeks it became a national frenzy. By the time the rescuers managed to reach Floyd, he had been dead three days and the press drifted away. The musical follows this storyline, opening with Floyd discovering a beautiful underground cave that he thinks will make his fortune only to get trapped on his way out. The rest of the musical is split between underground scenes in which Floyd's brother tries to talk him through the ordeal with childhood games of imagination, and the situation above ground where Floyd's family is joined by a rescue team and by national reporters hungry for a good story. In the final moments, a narrator informs the audience that by the time the rescue mission reached Floyd, he was already dead. The initial 1993 version of *Floyd Collins* at the American Music Theatre in Philadelphia was reworked and was followed three years later by a production at Playwrights Horizons. Since then, it has become one of the most acclaimed musicals to emerge from the American nonprofit theatre and the show has enjoyed numerous productions at major nonprofit theatres such as the Old Globe in San Diego, the American Music Theatre in Philadelphia (now renamed the Prince Music Theatre), and the Goodman Theatre in Chicago. The success of *Floyd Collins* can partly be traced to the original and sensitive way in which it explores some of the fundamental questions of contemporary American culture and identity through the music and libretto. It has also served as a new model for storytelling and staging, demonstrating how a story that had multiple settings and a large cast of characters could be successfully made into a small-scale musical and produced on a modest budget. The relative simplicity of the staging, while very much an aesthetic choice, has undoubtedly helped the afterlife of the show, enabling small venues to stage productions that do not come across as scaled-down versions of the original.

The structure of *Floyd Collins* and its ability to encompass broad ideas in a modest staging is to a great extent due to the fact that it was co-conceived, cowritten, and initially staged by an artist from an avant-garde rather than a traditional Broadway musical theatre background. *Floyd Collins* is most often associated with its composer-lyricist, Adam Guettel, rather than its writer-director Tina Landau. Guettel is deservedly acclaimed as an original and prodigiously talented musical theatre writer whose other work includes *Myths and Hymns* and *The Light in the Piazza*. He is the first to acknowledge that *Floyd Collins* was a collaborative writing project and that Tina Landau had an enormous impact on the finished show. Like James Lapine, Landau's background, interests, and aesthetics set her apart from traditional Broadway musical theatre directors. Although she has a long-standing fascination with classic musical theatre and has lately tackled more populist projects (including the 2001 Broadway revival of *Bells are Ringing*), Landau is primarily known for her work on more experimental theatre, finding her artistic home away from Broadway. In the summer of 1994, between the Philadelphia and New York mountings of *Floyd Collins*, she wrote and directed *Stonewall: Night Variations*. Depicting the start of gay liberation with the Stonewall riot in 1969, the piece was staged outdoors on Pier 25 by the Hudson River and included music, street theatre, and a cast of 60. Her musical theatre work has mainly consisted of psychologically driven collaborations with composer-lyricist Ricky Ian Gordon. This has resulted in pieces such as *States of Independence* (in which a suburban girl is visited in her sleep by a number of Revolutionary War figures) and *Dream True* (in which a boy is transplanted from his Wyoming home to live with a rich, troubled uncle and spends the rest of his life trying to undo the psychological damage). Other collaborations have seen her working with avant-garde director Anne Bogart (*American Vaudeville*) and undertaking a wide range of work at nonprofit theatres such as the Public Theatre, Playwrights Horizons, Chicago's Steppenwolf Theatre, La Jolla Playhouse, and Actors Theatre of Louisville.

This background informed both the development and staging of *Floyd Collins*, which reflects the nontraditional approach of its two creators. The main previous dramatization of this story, Billy Wilder's 1951 movie *Ace in the Hole*, focuses on the "carnival" angle through the central character of a cynical, alcoholic reporter (Kirk Douglas) who conspires with the sheriff to prolong the rescue operation from 16 hours to a week in order to create a news story. Guettel and Landau approached the story from both a psychological and sociological angle. While Guettel is highly knowledgeable about the Broadway canon, his own work has been in the nonprofit sector and his approach to *Floyd Collins* was based on exploring some rather personal psychological questions rather than creating a crowd-pleasing show. As a writer he necessarily spends a lot of time alone ("in my own metaphorical hole") and this drew him to the figure of Floyd in the cave.[26] In addition, while Guettel has been widely acclaimed as a musical theatre writer, the fact that he is the grandson of Richard Rodgers has always brought with it a certain burden of expectation and this attracted him to a story about evaluating our lives by the quality of the journey rather than by end results: "I wanted to work on a story where someone had aspirations for something beyond their reach, failed at attaining those goals and still had a noble life."[27] The emphasis on psychology and on the inner and outer life of the leading character underpins Guettel's score, which uses the Kentucky vernacular in both music and lyrics but also creates its own musical landscape, contrasting the freewheeling sounds below ground (including an extended, highly evocative yodel sequence) with more traditional folk sounds for the scenes above ground. Landau took a more sociological approach. In a subsequent interview, she explained that the musical originated from her desire to do a piece about "what it is to be American in this culture at this time."[28] In other words, the Floyd Collins story was a metaphor for American society and, as John Guare phrases it, "the paradox of what it's like to live in the most bountiful country in the world and at the same time be on the edge of an abyss that separates you from ever getting there."[29]

The richness of the final show (a mixture of mythical and prosaic, psychological and social, sophisticated and simple elements) can be partly traced back to the different points of entry of the two writers, but also to the collaboration that required them both to address all these different aspects of the story. According to Guettel, they worked closely together on all aspects of the writing, discussing everything at length and each contributing to the shaping of the whole:

> Every time we met we would talk about structure and the importance of those things. Some days we would meet and the societal strands versus family, personal, spiritual strands wouldn't come up; some days it was sheer structure working out how and what will happen in this particular scene and spotting the songs. We did everything together and she wrote the spoken words and I wrote the sung words. But in terms of weighing out for each of us what our homework was, what our assignments were, we would do that together and the tone of the piece would be developed during those conversations.[30]

Landau's dual role as writer and director on the project—and her theatrical background—undoubtedly affected what the audience saw onstage in the original production. Guettel maintains that their approach was to write first and then find a way of staging the result: "Most of the time we would allow ourselves to be connected to the emotional through line and worry about the practical matters when we got to them."[31] It seems questionable whether a more traditional musical theatre director would have felt comfortable writing such an epic piece in the knowledge that it would be staged on small stages with limited resources. On the surface, the multiple locations and the large crowds involved (family, community, press, tourists) would seem to call for a large theatre and sophisticated stage technology. Certainly, it is not hard to imagine a commercial version of this show in which the carnival of press, tourists, and locals triggers a rousing, festive dance number at the opening of Act Two. Neither is it a stretch to imagine what a commercial director and design team might have made of the underground scenes in the tunnels and caves.

The physical realization of *Floyd Collins*, however, came out of Landau's background in smaller-scale theatre where simple, nonliteral staging is often both a pragmatic necessity and an aesthetic choice. On *Floyd Collins*, the emphasis was firmly on psychological exploration, approaching the sociological issues from the characters' point of view rather than viewing them from the outside. Landau's staging emphasized the characters' state of mind rather than attempting an objective representation of the settings through painted drops or constructed scenery. Relying largely on lighting to set the mood, the stage became a psychological space with a simple and evocative set consisting of a screen across the back wall and a few wooden structures— ladders, planks, and boxes—on which characters leant and sat informally, creating a sense of their world without specifically recreating any one location (see figure 6.2). The underground scenes were staged simply on the same level as the rest of the action. For Floyd's opening travels, lights appeared in different parts of the stage to give a sense of his movement through space while the actor mimed squeezing through holes. Later in the show, Floyd's entrapment in the cave was represented by having the actor wedged against a simple wooden structure at down-stage right, with the various rescuers approaching him along an (imaginary) path from upstage right. Thus the above-ground scenes took place just a few feet away from the trapped Floyd.

This staging, as well as requiring the audience's imaginative collaboration, also helped to highlight the story's metaphorical themes of freedom and imprisonment. One of the key challenges of the story is that the main

Figure 6.2   Scene from *Floyd Collins*

character spends most of the show trapped in a cave unable to move. By staging the show in a nonliteral fashion, Landau was able to play with the idea of emotional and spiritual release. In her production, Floyd became physically free when engaged in flights of fantasy such as remembering childhood adventures with his brother. This also allowed for an ending that was inspiring rather than prosaic. The last moments of the show have an extraordinary emotional power and it comes from the fact that the death of Floyd is much less important than the transformative power of imagination and hope that turns his life into a rich journey that most of us can only aspire to. In Landau's production, just before the narrator announced the details of his death, Floyd was shown enjoying a moment of exquisite physical release as he imagined his escape from the mine; although he dies, the staging implies, his gift for seeing life in vivid and imaginative ways remained undefeated until the end.

Whether their shows infused new life on Broadway or remained in the nonprofit sector, writer-directors such as Wolfe and Landau have made a vital contribution to redefining the musical. By conceiving and developing musicals as offshoots of the more experimental and probing aesthetic of the nonprofit theatres, they and their collaborators have created new variations on the musical dramas initiated by Harold Prince on Broadway and developed through the West End musicals of the 1980s and 90s, combining the emotional power of music with the intellectual engagement of social and psychological dramas.

7. Rethinking Revivals ৵৹

So far, the focus of this book has been on the ways in which directors have helped to foster a more fluid relationship between British, American, subsidized, and commercial theatre to create new musical dramas. But this discussion can also be extended to recent revivals of classic musicals in which directors have treated the libretto and score as they would a play or opera, seeing them as living texts to be explored and reinterpreted rather than as museum pieces to be lovingly restored.

This shift in the musical theatre echoes a key development in both British and American drama in the twentieth century: namely, the shift in our relationship to classic texts as a result of the rise of the director. Historically, there is a long tradition of revisionist productions that includes the Roman adaptations of Greek plays, the reworking of ancient scripts by the neoclassicists, and the rewriting of Shakespeare plays by William Davenant and Nahum Tate for Restoration and eighteenth-century audiences—including Tate's infamous happy ending to *King Lear*. Prior to the twentieth century, these changes largely consisted of adapting the texts themselves to reflect shifts in cultural outlook or, more practically, to accommodate star performers.

From the late nineteenth century onward, the rise of the director as a central artistic figure increased the emphasis on overall production style. However, the rise of the conceptual director dates more precisely to modernism and the work of artists such as Gordon Craig, Vsevolod Meyerhold, Bertolt Brecht, Antonin Artaud, and Jerzy Grotowski: while these artists differed widely in philosophy and emphasis, the determining factor in all their work was the directorial vision. This resulted not only in original works (many of these artists were auteur directors) but also in new perspectives on classical plays through textual adaptations and interpretations. Examples include Craig's 1911 production of *Hamlet* at the Moscow Arts Theatre (in which Hamlet was a lone figure in black, walking through an all-white world) and Brecht's adaptation of John Gay's 1727 ballad opera

The Beggar's Opera into the more contemporary setting of *The Threepenny Opera* (1928).

The idea of the director as interpreter of classic plays, while originating earlier in the century, has intensified since the 1960s. In America, avant-garde directors such as Andrei Serban, Robert Woodruff, JoAnne Akalaitis, and Anne Bogart have advanced the idea of the conceptual director through using existing texts as a springboard for new works, either by placing a new frame on a play or by using fragments of the text interspersed with other texts and images. Amy Green refers to this approach as "rewrighting," a term that acknowledges the authorial role of the director and emphasizes the idea that "these directors craft or shape old scripts into new theatrical events."[1] The result has been a body of work that maps contemporary ideas and preoccupations onto older texts. As Green points out, "The classics have been made to absorb and/or confront the issues and aesthetics of America's turbulent, fragmented, and cynical century. Through these controversial productions, we have revised our view of the plays and they, in turn, have reflected images of us."[2] In Britain, the newly created institutional theatres created a very different kind of crucible for exploring the dramatic canon, not least the rejuvenated Shakespeare Memorial Theatre in Stratford-upon-Avon. With the arrival of Peter Hall as Artistic Director in 1960, the theatre was reborn as the Royal Shakespeare Company (RSC) and entered a period of exploratory productions conceived and directed by university graduates trained in textual analysis and the classics. During the 1960s and 70s, directors such as Peter Hall, Trevor Nunn, Peter Brook, Terry Hands, and John Barton started to create a production canon that offered striking new interpretations of the classics with close adherence to the written text.

THE MUSICAL REVIVAL

While the twentieth century has seen an increased engagement with classic plays, the musical revival was relatively unheard of prior to the 1980s. This is partly due to the youth of the genre: unlike drama and opera, musicals are essentially a twentieth-century art form. Furthermore, many of the early shows (revues, musical comedies) were intended as light topical entertainment whose characters and plots have not stood the test of time; generally speaking it is the more dramaturgically and thematically sophisticated musicals of the post-*Oklahoma!* era that lend themselves to revival. As a result, there has been little time or perceived need for directorial reinterpretation and the common practice prior to the 1980s was to do out-of-town tryouts, run a musical on Broadway for as long as there was an audience, tour the show nationally and then possibly bring back the same production to Broadway if there was sufficient demand. Occasionally (as with *Hello, Dolly!*) the original production would be brought back for a limited run after an interval of a few years and was usually directed by somebody involved in the original production. In his book *Everything Was Possible: The Birth of the Musical Follies*, Ted Chapin quotes an interview with Gene Nelson, one of the stars of *Follies* who subsequently directed a revival of the show. Faced with actors wanting to discuss the show, Nelson was uninterested: "I told them: This is the way we did it; this is the way you're going to do it. Don't talk to me about motivation."[3]

Starting in the 1980s, there has been a dramatic increase in musical revivals on Broadway. But while the numbers have increased, the role of the director has generally remained that of physical re-creator rather than textual interpreter. This may be partly explained by the fact that these revivals were usually nostalgic and escapist in purpose. On a practical level, the revivals were a response to three simultaneous pressures: the changing economics as ticket prices started to catch up with escalating production costs; the devastating effect of the AIDS epidemic on the artistic community; and the sudden loss of the directors who had been the driving force behind new musicals since the 1960s.[4] While the West End was experiencing a burst of creativity, Broadway dwelt on its past achievements with revivals of shows such as *Cabaret, She Loves Me, Man of La Mancha, A Funny Thing Happened on the Way to the Forum, Chicago, Grease,* and *Guys and Dolls* as well as artist retrospectives and collections of past hits in revue format (*Smokey Joe's Café*) or repackaged as "new" book musicals (*My One and Only, Crazy For You, Ain't Misbehavin*). The aim of these was overwhelmingly to celebrate the past rather than to interrogate classic texts from a contemporary perspective.

AFTERLIFE

The very term "musical revival" reveals a significant difference between how we treat the musical theatre canon and other performing arts: classic operas and plays receive new "productions" but musicals are "revived." The question of alternative approaches has not yet been a pressing issue due to the relative youth of the art form, but it is one that needs to be addressed if the canon is to continue to occupy a vibrant place in our cultural lives. The central argument in Jonathan Miller's *Subsequent Performances* provides a useful reference point in this respect. While Miller's focus is primarily on Shakespeare productions, his discussion of the "afterlife" of a work applies to all art:[5]

> The destiny of a great play is to undergo a series of performances each of which is incomplete, and in some cases may prove misleading and perverse. By submitting itself to the possibility of successive re-creation, however, the play passes through the development that is its birthright, and its meaning begins to be fully appreciated only when it enters a period that I shall call its afterlife.[6]

Drawing historical parallels, Miller uses the term as a way of highlighting the changing meanings of a work of art with the passage of time: "There comes a point in the life of any cultural artifact, whether a play or a painting, when the continued existence of the physical token that represents it does not necessarily mean the original identity of the work survives."[7] The relevance of this idea to the performing arts seems clear. Social attitudes and preoccupations change over time as do performance styles and if a show is to feel like anything other than a museum piece then the director must have the ability to approach it from a new standpoint. To attempt a faithful re-creation of the original staging may preserve the physical shell, but it does not re-create the audience's experience at being confronted by the work. As Miller argues,

> The preservation of the canonical performance, which many people think would express and embody Shakespeare's intentions, is more likely to cut us off from his vital dramatic imagination as we would be distracted by the foreign gestures and styles of acting. It seems to me that it is precisely because subsequent performances of Shakespeare's plays are interpretations, rather than copies, that they have survived.[8]

While our separation from classic musicals is measured in decades rather than centuries, this point still holds true for musical revivals. If theatrical meaning is created by the interaction of art and audiences, and if an audience's outlook is at least partially a product of their time, then it follows that even an impeccably reproduced production will be a very different experience for a contemporary audience than it was even 10 or 20 years earlier. What was innovative seems quaint; what was contemporary looks dated. It seems clear, then, that if the classic musicals are to remain culturally relevant then directors must be allowed to approach these texts anew rather than simply re-creating their predecessors' staging.

"SHOW" AND "PRODUCTION"

To effect this change with musical revivals we need to confront one of the underlying problems with the traditional approach: the conflation of the show with its original production. Again, this has precedents in drama where groundbreaking twentieth-century dramatists such as Chekhov and Brecht are often very closely linked with the original productions of their work. In such cases, as Miller points out, "The colleagues who were closely associated with his pioneering efforts often appoint themselves as custodians of the orthodoxy and ensure that all the remembered details of the inaugural production are preserved from one performance to the next."[9] This phenomenon has certainly underpinned the approach to the musical theatre canon, with revivals often being overseen by original members of the cast or crew who attempt to re-create the original staging. However, if the musical canon is to enjoy an interesting afterlife it is essential that we recognize the difference between the show (music, lyrics, dialogue, and their underlying themes) and the original production (the physical realization of these elements). In other words, we need to acknowledge that the original production is just one of several possible embodiments of the show itself as defined by the libretto and score.

However, while this distinction can and should be made, it would be naïve to ignore the complications that arise from the fact that the musical is a highly collaborative art form. More than plays or operas, musicals are partly created in the rehearsal room and in previews, with the writers, director, choreographer, and designers often participating more in the creative process. Given this, the original "interpretive" creative team (director, choreographer, designers) can have an impact on the work that ends up not just on the stage but in the published play script, such as Prince's visit to the Taganka theatre, which resulted in a complete restructuring of the opening number in *Cabaret*. Similarly, the use of a revolve at the Barbican Theatre created the possibility of fluid transitions that helped to shape the dramatic structure of *Les Misérables* as a text rather than just the original production. The difficulty of separating the show from the original production is intensified when dealing with the choreographer-director driven shows from the 1960s and 70s. The staging and choreography of Jerome Robbins, Gower Champion, Michael Bennett, and Bob Fosse are widely considered an integral part of their shows. Certainly it could be argued that the most abiding memory from *A Chorus Line* is not the characters or the story, but Bennett's choreography including the fluid breaking and reforming of the line of dancers at the front of the stage; similarly, the lingering image of *Sweet*

Charity is of the decadent line of hostesses leaning over a metal bar as they languidly plied their trade, and of the angular, disdainful poses in "Rich Man's Frug." Indeed, the centrality of the staging in Bennett and Fosse musicals may be one reason why they are attempted less often than musical plays built around a strong libretto and more complicated themes. However, it does not follow that their staging is an integral part of the show itself. It is one thing to claim that Robbins functioned as an author (or as a dramaturg) of *Fiddler on the Roof* when he told the writers to come up with a unifying theme for the show, resulting in the opening number "Tradition." But it is quite another to argue that his staging of that number is integral to the show itself. Equally, Fosse's taste for chic sleaze, show business, and decadence have helped to establish the tone of *Sweet Charity* and *Chicago*, but it does not follow that his distinctive choreographic vocabulary are also part of the "text" of the shows themselves.

In the 1990s, the problematic idea of the original production as an integral part of the show was increasingly challenged as high-profile new interpretations began to emerge. Just as new musical dramas challenged the traditional models of writing and staging, so these new productions of classic Broadway musicals questioned the idea of the "revival" as a reenactment of the original production. Rather than being exercises in nostalgia, these productions offered fresh interpretations, opening up new possibilities for how to approach the musical theatre canon. Interestingly, while many of these productions reached Broadway, they largely originated elsewhere. Hal Prince's revival of *Show Boat* was developed in Toronto by Canadian producer Garth Drabinsky and played in the West End before reaching Broadway in 1994. Encompassing songs from different stage versions and the film version of the show, the most radical innovation was the creation of a new Act Two opening number to address the racial stereotypes in the original. In 1996, Christopher Renshaw's Australian production of *The King and I* arrived on Broadway, with a much greater emphasis on the culture clash inherent in the show and a more authentic portrayal of "Siam" and the "Siamese" characters, although Jerome Robbins's original choreography was retained. Away from Broadway, American directors also offered some striking new interpretations of classic musicals. In 1984, Anne Bogart's production of *South Pacific* at New York University reconceived the show as a piece enacted by inmates of an asylum during World War II against the backdrop of news from the Front. More recently, Amanda Dehnert's 2003 production of *Annie* at Trinity Rep pointed out the bleaker side of the story by emphasizing the homeless through staging reminiscent of the Brecht-Weill collaborations. The intention was to have Annie waking up back at her orphanage at the end, the whole thing just a dream—a twist that was abandoned when the librettist, Martin Charnin, exercised his authorial veto.

BRITISH DIRECTORS AND
MUSICAL REVIVALS

But despite these isolated attempts at reinterpretation, the most revelatory "Broadway" revivals have arguably been a group of British-directed productions. This can possibly be explained by three cultural factors: the directors' relationship to the original productions; their experience with large-scale classical theatre; and the general trend for reinvestigating the classics that characterized London theatre in the 1990s. The geographical and cultural distance of the British directors from the American musical is undoubtedly the key factor here. While Broadway musicals have a strong following in the United Kingdom there is not the same cultural attachment to the original staging as there is in America. When directing *Follies* on Broadway in 2001, Matthew Warchus was struck by "just how many people had a sense of ownership in New York" adding that when it comes to Broadway musicals, "there's no real sense that these things are up for grabs and ready for refreshing. There's a sense that this is hallowed ground."[10] By contrast, as Jack Viertel points out, the British directors are "liberated to do these productions by the fact that they don't have any stake or ownership in what was done in the past."[11]

The implications of this cultural difference are exemplified by the 2004 Broadway revival of *Fiddler on the Roof* directed by David Leveaux for Roundabout Theatre Company. The musical, which premiered on Broadway in 1964, depicts a tight-knit community of Russian Jews in the era of the pogroms and follows the fortunes of Tevye the milkman, his wife, and daughters as their family is threatened not only by the political situation and the Russian pogroms but also by the younger generation's new ideas concerning marriage and community. For the Jewish-American creative team, the show was a very personal tribute to their ancestors and over the past 40 years *Fiddler* has become best known as the first mainstream musical by, about (and, given the Broadway demographics, to a great extent *for*) American Jews. This cultural association was clear in the original production, which drew on traditions of American musical comedy and Yiddish theatre, with Zero Mostel—the original Tevye—frequently ad-libbing Yiddish phrases and jokes during performances. From the vocal delivery to the dance configurations and the music, the original production of *Fiddler on the Roof*, directed by Jerome Robbins, was as much a celebration and affirmation of Jewish-American culture as an elegy for Russian Jews in a nineteenth-century shtetl. This specific set of meanings and associations has since been perpetuated in the show's productions in regional theatres, school productions, and Jewish cultural centers across America.

Rather than offering a Broadway-centric depiction of Russian Jews filtered through Jewish-American culture, Leveaux's production was a different response to the show's setting and to the underlying themes of tradition and family as written in the libretto and score. The central performance by Alfred Molina was more restrained than that of the extrovert Mostel and the set was more sparse and elegiac than is customary for this show, mainly consisting of real trees scattered over a bare stage. As a result, the design was more reminiscent of a Chekhov set than a traditional production of *Fiddler on the Roof*. The critical and audience response to Leveaux's productions was divided and the production has also stimulated debate in academic circles, most notably in Jessica Hillman's article "Goyim on the Roof: Embodying Authenticity in Leveaux's *Fiddler on the Roof*." It is not my intention here to engage in a critical debate over the specific staging and interpretive choices of this production, but it does serve to illustrate the way in which critical evaluations of musical revivals are often based on the cultural associations and sense of ownership that have grown up around a show rather than an evaluation of how the new production sheds new light on a well-known text. I would argue that, like all good drama, *Fiddler on the Roof* lends itself to multiple readings. In order to reinterpret the piece, however, we need to acknowledge that the original production was itself an interpretation of the underlying text about family, community, and traditions. However magically it was originally staged, it seems to me unhelpful to hold this up as the only possible production rather than exploring new and unexpected readings of the show. Certainly the writers themselves do not appear to view the first production as definitive and seem open to multiple readings. At a panel discussion in 2003, the librettist, lyricist, and composer spoke of their dismay at Mostel's insertion of Yiddish improvisations, which broke the illusion of characters who were all speaking in the same language all the time; librettist Joseph Stein also remembered being touched at the Tokyo opening where a theatregoer asked him how American writers could have such a deep understanding of the inner workings of Japanese family life.[12]

In addition to having a certain cultural and geographic distance from the shows, the British directors' approach to classic musicals was also in some measure due to their background in British institutional theatre. The 1980s saw the emergence of a generation of theatre directors whose work on classical dramas in the large subsidized theatres gave them experience of textual analysis and of large-scale productions that, like the Golden Age musicals, required the ability to handle large casts and theatre spaces. Todd Haimes is the artistic director of the nonprofit Roundabout Theatre Company in New York, which produced three British-directed musical revivals on Broadway: Sam Mendes's *Cabaret*, Matthew Warchus's *Follies*, and David Leveaux's *Nine*. His reasoning for the preponderance of British musical theatre directors on Broadway is the fact that young British directors have early access to the large stages and impressive casts of the National Theatre or the RSC. He points out that "there's no American equivalent for that kind of experience. When you're a hot young director in London you direct Derek Jacobi at the National in Shakespeare; here, hot young directors will do a new play at Manhattan Theatre Club with five characters."[13]

Finally, the sudden spate of new musical revivals was part of a general trend for classical reexaminations in London theatre in the 1980s and 90s— a phenomenon that might in part be ascribed to a *fin-de-siècle* desire for social and cultural reassessment. During the 1980s, the English National Opera presented reinvestigations of opera favorites and in 1992, Stephen Daldry's National Theatre production of J.B. Priestley's *An Inspector Calls* turned a seemingly stodgy mid-century morality play into a devastating indictment of the Thatcher era. In addition, reinterpreting the classics was the main emphasis of three of the most prominent British theatre companies in the 1990s: Cheek by Jowl, which toured nationally and abroad with vigorous ensemble-based productions of Shakespeare; the Almeida Theatre Company, which offered muscular new productions of European classics; and the Donmar Warehouse, which built a reputation for fresh productions of modern American classics. The breakthrough in musical theatre revivals lay largely in the application of this investigative approach to the Broadway classics by British directors such as Nicholas Hytner, Trevor Nunn, Sam Mendes, Matthew Warchus, and David Leveaux. These directors have openly challenged the idea of the musical as fundamentally different from other theatrical genre. Sam Mendes has argued that "*Cabaret* is up there with *The Crucible* or *The Homecoming* or any other great play of the twentieth century that deserves to be reinvented and rediscovered generation to generation: it's a great piece of theatre."[14] David Leveaux concurs, pointing out that directing plays and musicals requires some of the same fundamental skills: "The truth is that directing *Betrayal* is absolutely a function of rhythm: inner rhythm. Directing a musical: absolutely a function of rhythm."[15]

Rather than re-creating the original productions, these directors based their interpretations on the "book" in the broadest sense of the characters, libretto, score, and themes. Their background in close textual analysis prompted them to mine the texts for new meanings, to think thematically about the shows and to place them in specific social and historical contexts. It was presumably this rigorous textual background (coupled with the fact that the productions would happen 3,000 miles from Broadway) that persuaded the Rodgers and Hammerstein Organization to give them unusual freedom with *Carousel* (directed by Nicholas Hytner) and *Oklahoma!* (directed by Trevor Nunn) including unprecedented permission to replace the original choreography by Agnes de Mille in both shows.

There are, naturally, limitations to this approach: in particular, the thematic and textual approach is less useful with the more extrovert demands of musical comedy. Matthew Warchus points out:

> What Broadway does fantastically well is this thing of energy and "presentation"— selling a number—which in Europe people don't do so easily. If you've got a show that requires that then it can be tricky in Europe whereas it is second nature for Broadway performers. . . . Maybe it's connected to the American assertiveness and confidence—you know, "I'm here and this is what I can do." Whereas there's something more shielded and circumlocutory and protective and subtextual about British culture and maybe other European cultures as well. And maybe that is somewhat manifested in performance energy.[16]

In addition, Jack Viertel notes that there is a distinct difference between British and American staging traditions and that this can be both an advantage and a disadvantage when tackling shows that were created in a different theatrical culture:

> They're a little hard to quantify but there are stylistic tics that are popular there [in England] in the same way that there are here. . . . It was interesting watching Nicholas Hytner's *Carousel*, for example. It didn't look anything like an American *Carousel* and it didn't deal with the way you get from one place to another like a typical American production. But in *Carousel* you don't have to. Other shows are very hard to do unless you use the ride-out in a certain way. They're written to be done that way.[17]

But whatever the limitations or criticisms of individual productions— and valid criticisms can of course always be made—there is no question that this new approach has helped to change the way that we think about the classics. In chapter 8, I offer a closer examination of five British-directed musical "revivals." Like new musical dramas, these productions have applied ideas from other performing arts, inviting us to view musicals as we do well-known plays and operas and paving the way for a more active and demanding relationship between contemporary audiences and the musical theatre canon.

8. Staging the Canon: British Directors and Classic American Musicals ∽

The new approach to the musical canon discussed in chapter seven has wrought a significant change in how we think about the classics. In this chapter, I look specifically at the way in which five British-directed productions have helped to redefine our relationship with classic musicals: Nicholas Hytner's *Carousel*, Trevor Nunn's *Oklahoma!*, Sam Mendes's *Cabaret*, Matthew Warchus's *Follies*, and David Leveaux's *Nine*. All five productions originated in London or New York and they all played on Broadway between 1993 and 2004. It is not my contention that these were the first or the only attempts to look at classic musicals through fresh eyes. But the high profile of these productions and the critical interest that they have stimulated has helped to normalize a more interpretive approach to the musical canon. Most importantly in terms of this book, they have intensified the connections between mainstream musical theatre and drama by emphasizing new dramaturgical and thematic approaches to the material and, crucially, physical staging inspired by the text rather than the original production.

NICHOLAS HYTNER'S *CAROUSEL*

Nicholas Hytner's 1992 National Theatre production of *Carousel* (which transferred to New York's Vivian Beaumont Theatre in 1994) is a particularly strong example of the need to engage critically with musical theatre classics. Based on Ferenc Molnar's play *Lilliom*, *Carousel* (1945) tells the story of an unlikely romance between Julie Jordan, a factory worker, and Billy Bigelow, an unrefined fairground barker. They marry after an awkward courtship, and Julie stays faithful to Billy even when he expresses his frustrations with life by hitting her. Billy dies in a robbery he undertakes to make money for his unborn child, and Julie is left to bring up their daughter Louise. In a scene set 16 years later, Billy is allowed down to earth for a day to make amends and seeks out Louise: he sees her on the beach with wild teenage boys, watches her dance with one of them, and later strikes her in a fit of temper. The show ends with Louise's graduation where Louise and Julie feel Billy's presence and everyone joins in a reprisal of "You'll Never Walk Alone."

Productions of *Carousel* have traditionally emphasized the show's bucolic setting, the romance of the ballads, and the musical comedy moments, mainly provided by the courtship and marriage of Carrie Pepperidge to the capitalist, bourgeois Enoch Snow. As Frank Rich noted in his article on the Hytner production, "Hollywood and a thousand stock productions bowdlerized the show to fit the treacly conformist culture of a decade whose rigid dogma was sexless suburban family bliss."[1] It is no accident that the best-known song from the show (and the one that is reprised at the end) is the soaring ballad providing hope and comfort to both the characters and the audience. By contrast, Hytner confronted the show as a psychological drama that offers insights into the underlying loneliness and violence of the characters' lives. While the musical has softened Molnar's play considerably, it still has implicit dark undercurrents in the tensions between characters and in the depiction of their society. There is a sense of isolation in the four lead characters, none of whom completely connect with their respective partners. Through his staging of key moments, characterizations, and his choice of set designer and choreographer, Hytner emphasized the darker underlying themes of violence and loneliness that have traditionally been glossed over in favor of a more quaint evocation of the characters and events. As David Richards noted in the *New York Times*, this approach created a rich, disturbing emotional landscape: "By darkening the characters in this fashion, Mr. Hytner allows his actors to work with a denser subtext than usual. Romance can turn rough and sweaty. Quick tempers keep undermining the Puritan proprieties. . . . On every front, a heightened dialectic between light and shadow, decency and prurience is central to the vision and gives the production its distinction."[2]

Most obviously, Hytner revealed the implicit connection between sex and violence. One of the most problematic elements in the show is Billy's behavior toward his wife. In the libretto, this is not directly confronted and Julie explains away her husband's violence as a sign of his frustration and unhappiness. Equally, when Billy strikes Louise, she asks Julie why the slap did not hurt, to which her mother replies that it is possible for someone to hit you hard without actually hurting you. The implication is that because

Billy fundamentally loves Julie his violence is somehow permissible or at least bearable. A contemporary audience will recognize this as a classic case of domestic violence but in 1945 the connection would have been more opaque. This is not to suggest that these problems did not exist, but simply that our attitudes toward them have. In a newspaper interview, Hytner highlighted the challenge of tackling this kind of material 50 years later: "Not only can we address these topics differently, but in some cases—as with the domestic abuse in *Carousel*—changed social norms demand that we do so. The vilest thing you could do to it would be to present it as a sentimental, foot-tapping Broadway show because that would be to exploit these issues."[3] When Billy hit Louise in this production, it was a hard slap in the face. Equally, Louise's *pas de deux* on the beach was reconceived to bring out an angrier and more violent undercurrent of sexual tension. Agnes de Mille's 1943 choreography (revolutionary at the time for its sensuality) has traditionally been reproduced in revivals. Hytner brought in Kenneth MacMillan, the prestigious British ballet choreographer whose choice of complex and sometimes painful subject matter helped to push back the boundaries of classical dance, emphasizing psychological truth and emotional complexity over pretty steps. His new choreography for the *pas de deux* between Louise and a teenage boy on the beach evoked the raw emotions of two teenagers at odds with the world and prompted the *Variety* critic to observe that "de Mille choreographed a scene's superego, while MacMillan reveals its id."[4] Hammerstein's depiction of Billy's behavior was daring in the 1940s, but Hytner's more uncompromising depiction of the violence reflected a shift in social perspective, fulfilling Hammerstein's underlying aim of confronting audiences with contemporary social issues.

The overall design of the show was also a result of Hytner's more psychological approach to the material, reflecting the underlying loneliness of the characters by depicting the isolated landscape around Maine rather than evoking the cozy, idealized Americana of more traditional productions. Designer Bob Crowley, who conducted research in Maine, has stated that his inspiration came more from the landscape's sense of vastness than from the domesticated, romanticized setting that has become part of the show's image: "A lot of the time you do get the figure of the actor against a lot of space. That's what I felt about the coast of Maine—there's the whole huge Atlantic in front of you and the whole of America behind you."[5] The result, as Frank Rich noted, was "more Edward Hopper than Norman Rockwell"—a world in which the characters "grope desperately for each other against a vast moonlit night that only emphasizes their lowly status in an indifferent universe."[6] The emphasis on involvement in the inner lives of the characters was also aided by the use of a central revolve. While the revolve is not a new technical innovation, it is not usually employed in productions of *Carousel*, which tend to rely on flats and backdrops painted with bucolic images. The decision to create the world of the show with a revolve and lighting projections on the back wall helped to give a sense of open space and also enabled more fluid transitions between scenes, emphasizing the ongoing story rather than the more distracting practice of scenery sliding and flying on and off the stage.

Hytner's social and psychological readings of *Carousel* were reflected even in the opening and closing moments of his production. *Carousel* traditionally opens with the revelation of a fairground carousel: while Julie mentions that she works in a mill, the audience's visual frame of reference is all bright lights and cheery noise. Hytner, however, emphasized the everyday horrors of working-class life and, in the process, offered a further explanation of why Julie is so fascinated by Billy that she misses her curfew to spend time with him on their first meeting. Instead of starting at the fairground, Hytner raised the curtain on the inside of an oppressive cotton mill with a row of factory workers sitting like automatons in front of an enormous, dominating clock ticking out the minutes until their release from captivity (see figure 8.1). When the hands of the clock finally reached six o'clock, the mill workers clocked out and passed through the gates and only then did the

Figure 8.1   Opening scene from Nicholas Hytner's *Carousel*

fairground scene appear. The use of the revolve was crucial to Hytner's vision of the opening. As Jane Edwardes noted in London's *Time Out*:

> One scene seamlessly flows into another as the girls working in the cotton mill finish for the day, join the fishermen swinging planks of wood as though they are as light as skipping ropes, and flock to savor the bearded ladies and bears at the fair. The gaudy carousel of the title is gradually put together until, as Rodgers' edgy waltz gets into its stride, its top breathtakingly unfolds like a parasol and the horses spin.[7]

Through this continuous movement, the audience was able to experience the arrival at the fairground through the characters' eyes, moving from their bleak working lives through the streets to the bright lights, music, and movement of the carousel.

Hytner's emphasis on the essential isolation of the characters in the show resulted in a radical final moment. Traditionally, the finale offers a sense of resolution and redemption with Billy having reestablished himself (even as a ghost) as a positive force in Julie and Louise's lives and given them a sense of hope: the reprisal of "You'll Never Walk Alone" by the entire cast creates a final sense of optimism. In Hytner's production, the song was accompanied by the image of Billy ascending a staircase back to heaven—the outcast figure who, although somewhat redeemed, is still very much alone. This final moment offers a telling commentary on how our perspective has shifted since 1945, a fact that Frank Rich pointed out at the time:

> As Billy has an afterlife, so does the America with which Rodgers and Hammerstein surrounded him. . . . Everything about this musical says we are alone. The audience crying at *Carousel* realizes that it is up to us to end this country's unending cycle of social injustice and domestic violence. And that not even Rodgers and Hammerstein, the soothing parental figures we had always depended upon can bail us out.[8]

As with the other changes in the production, this final moment did not involve textual changes but rather a shift of emphasis through a new physical realization of the libretto.

TREVOR NUNN'S *OKLAHOMA!*

Where Hytner offered a psychological reading of *Carousel*, Trevor Nunn's 1998 National Theatre production of *Oklahoma!* had a more sociological emphasis. Like Lynn Riggs's play *Green Grow the Lilacs* on which the musical is based, Nunn's production focused on the everyday struggles of people living on the frontier.[9] His reading of the show was as a social drama about a crucial moment in American history when the expansion into the West necessitated a sometimes strained coexistence between different kinds of people: "These people find themselves settling in God's acre, in a kind of paradise. They must decide who is going to live there and on what terms. It's a turbulent question. There's a kind of war going on, a war about fences between cowmen and farmers. It's a nervy time."[10] Nunn's emphasis on the uncertainty of the period reflects Andrea Most's analysis of the show's implicit social tensions. In *Making Americans: Jews and the Broadway Musical*, Most highlights the racial and ethnic imagery around the show's two "outsiders": the Jewish characteristics of the "Persian" peddler Ali Hakim and the African-American connotations in how the brooding, sometimes violent farmhand Jud is described and portrayed.[11] While Nunn's production is not racially oriented, it does echo Most's emphasis on a world in which Jud and Ali's unorthodox attitudes constitute a threat to a fragile social order.

Oklahoma! has acquired a nostalgic association over the years due to the tradition of quaint sets and costumes and to the optimism in songs like "Oh What a Beautiful Morning," "The Surrey with a Fringe On Top," and the rousing title song, all of which celebrate an idealized American West. In Max Wilk's *OK! The Story of Oklahoma!*, the original costume designer Miles White recalls that set designer Lemuel Ayers "tried for a spare and vivid look based on Grandma Moses images,"[12] referring to the work of Anna Mary Robertson "Grandma" Moses whose lively, colorful depictions of America's rural past (sleigh rides, candle making, sheep shearing, harvesting hay) became widely popular after her work was discovered in 1938. This quaint depiction of the show has been crystallized through the 1955 Hollywood film and the show's frequent revival by schools, amateur groups, and professional theatres that have established *Oklahoma!* as a symbol of nostalgia in the popular American imagination. In 1998, Susan Stroman—the American choreographer of Nunn's production—noted wryly of the set that "if this were the usual American production, that house would have geraniums all over it."[13]

Nunn deliberately distanced himself from this tradition. Stroman has pointed out that while the show is "revered, sacred ground" in America, they approached it "as if it were a new work."[14] Nunn's reasoning was that we no longer look at the settling of the American West in the same way. Just as Clint Eastwood and Robert Altman have changed the idea of the Western movie, he argued, so we need to revise our relationship with *Oklahoma!*: "The film's location was thrilling. But back then naturalism didn't have to include crumpled clothing, dirt, sweat and real conditions. These people live out beyond civilization and the law. They're in a territory, trying to work things out for themselves. Yet in the film there's a sense they all have a fridge and a telephone in their homes."[15] Nunn's more contemporary, sociological perspective on the taming of the West was reflected in designer Anthony Ward's set design. As in *Carousel* (and previously in Nunn and Caird's *Les Misérables*) the staging of this *Oklahoma!* was based on a large central revolve and made creative use of lighting on the back wall to evoke the sense of open spaces. Aunt Eller's house was a simple structure and ensemble scenes conveyed the sense of a community who lived a basic and unglamorous life (see figure 8.2). The surrounding landscape was also suggested through representations of oddly slanted farmhouses and windmills and through a witty use of perspective such as towering sheaves of corn and a small model train that ran across the back of the stage. These were all set against the vast skies of the prairie that conveyed, in Michael Billington's words, "the territory's aching emptiness."[16] The emphasis on the open space rather than on a cluttered stage also allowed for a final social commentary as the happy couple drove off into the unforgiving dust bowl.

The more realistic depiction of the show's events and characters is reflected in the interactions between Laurey, Curly, and Jud. Where Curly is traditionally the hero and Jud the villain, both were played with more complexity (by Hugh Jackman and Shuler Hensley respectively) blurring

Figure 8.2   Company scene from Trevor Nunn's *Oklahoma!*

the distinctions between the two, problematizing Jud's social exclusion and creating a more genuine dilemma for Laurey (Josefina Gabrielle) as to who should accompany her to the dance. This struggle was intensified through the new choreography for the Dream Ballet, "Laurey Makes her Mind Up," in which the heroine's attraction to both Jud and Curly is enacted and her simultaneous fear and excitement over Jud is allowed to be expressed. The original choreography was created for dancers who replaced the principals during the number—common practice in 1943 when specialization was more pronounced and singing and dancing choruses had only recently been integrated. In Nunn's production, Stroman choreographed the number to be performed by the actors playing Laurey, Jud, and Curly, allowing the Dream Ballet to create a more intense evocation of Laurey's mental state in the dream and to avoid the distancing effect on the audience of having strangers assume the roles. Like MacMillan's new choreography in *Carousel*, this staging was a manifestation of the thematic directorial vision; it was, furthermore, a direct result of allowing the director's reading of the text to dictate the physical production and not vice versa.

SAM MENDES'S *CABARET*

Where Hytner and Nunn dealt with changing social contexts from 60 years earlier, Sam Mendes's production of the more recent *Cabaret* exemplified Jonathan Miller's idea that "in the last part of the twentieth century even the comparatively recent past is visualized as a foreign country where people do things differently."[17] Initially staged at the Donmar Warehouse in 1993, the production opened on Broadway in 1998 (with new choreography by Rob Marshall who was also credited as codirector) and incorporated songs that had been used in previous theatre productions and the 1971 film version. It was, like Hytner and Nunn's productions, an interpretation that was faithful to the themes and intentions of the show as written rather than the original staging.

Mendes's challenge was threefold: to find new ways of conveying the central metaphor, to find contemporary parallels for Prince's social allusions, and to address our changed perspective on the historical period of the show. At heart, both productions were concerned with the same underlying sociopolitical preoccupation: namely, the insidious spread of prejudice and evil as epitomized by the rise of Adolf Hitler and the Third Reich. Like Prince, Mendes took care to implicate his audience in the events onstage. Where Prince had literally held up a mirror to the audience, Mendes transformed the auditorium into the Kit Kat Klub with tables and waitress service. Prince's rationale for the show was echoed in Mendes's vision: "The space operate[s] as a metaphor for Germany. . . . In the early 1930's you were in a club having a great time. By the mid-30's, the door was being locked from the outside, and by 1939 you couldn't get out. It's a physical metaphor: you're not an observer; you're part of it; you're there."[18] Where Prince's Emcee lured in the audience through show business charm and panache, Mendes's production opened with a tiny spotlight on the door from which the Emcee's hand appeared, his finger beckoning the audience.

Mendes's production also reinvented Prince's idea of the show as a mirror of contemporary social developments. Where Prince used the production to highlight the prejudice and bigotry in Civil Rights-era America, resonating with contemporary playwrights, Mendes's production reflected the more rebellious, nihilistic, and violent tone of British popular culture 30 years later. In the 1990s, there was a wave of new plays in London by writers such as Mark Ravenhill (*Shopping and Fucking*), Sarah Kane (*Blasted*), and Martin McDonagh (*The Beauty Queen of Leenane*) who used violence, sex, and black humor to portray the sense of betrayal and nihilism in the generation raised under a prime minister who famously claimed that there was no such thing as society. In Mendes's *Cabaret*, he referenced heroin chic through the visibly drug-addled chorus girls and his staging was more consistently seedy and sinister than Prince's, evoking a darker reading of the underlying themes.

The staging of the title song in the two productions provides a clear illustration of these differences. At this point in the show Sally is living in denial of the political events around her; having just undergone an abortion and rejected the chance of a new life with Cliff, she returns to perform at the Kit Kat Klub. In the Prince staging, the song was a somewhat defiant and triumphant eleven o'clock number. Critic Walter Kerr noted at the time: "The décor says she has won. . . . It would take a very great actress indeed to give the lie to this stunning piece of design, and a most remarkable singer to find—in the melodic fullness of the title song—an implication that the character has made a very sorry choice."[19] This triumphant interpretation was intensified in Bob Fosse's film version in which Liza Minelli made the song into a rousing anthem to the triumph of hope over experience, evoking more admiration than pity for the character. By contrast, Sally's sorry choice was clearly underlined in the Mendes production where she performed the whole song on an eerily empty stage with just an old-fashioned, freestanding microphone in front of her. Here, Sally was not so much a triumphant figure as a tragic victim of her own self-destructive impulses. As

Ben Brantley noted in his *New York Times* review, this Sally

> isn't selling the song; she's selling the character ... this avidly ambitious chanteuse recoils when the glare hits her, flinching and raising a hand to shade her face. Wearing the barest of little black dresses and her eyes shimmering with fever, she looks raw, brutalized and helplessly exposed. And now she's going to sing us a song, an anthem to hedonism, about how life is a cabaret, old chum. She might as well be inviting you to hell.[20]

The depiction of sexuality in the Prince and Mendes productions is also markedly different. While Sally's blatant use of sex to advance her career is clear in the original production, Mendes's production took a more explicit and playfully aggressive approach to sexuality. In "Two Ladies," one of the chorus "girls" was a man and the three performers mimed vigorous sexual acts in silhouette behind a screen. In "Don't Tell Mama," Sally Bowles sat in an enormous chair dressed in a girlish dress—an echo of the controversial Calvin Klein advertisements featuring young models in suggestive poses. The most noticeable difference in portrayals of sexuality, however, was in the open depiction not only of Cliff's bisexuality but also the Emcee's—a difference that can be largely credited to a shift in attitudes to homosexuality between 1966 and 1993. When *Cabaret* first opened on Broadway it was pre-Stonewall, homosexuality was still a taboo subject and the persecution of homosexuals under the Nazis was not commonly acknowledged. In the intervening years, however, this connection was explicitly examined in plays such as Martin Sherman's *Bent* (1979), which highlighted the persecution of homosexuals under the Nazi regime through a love story set in a concentration camp, and Hugh Whitemore's *Cracking the Code* (1986), which told the story of Alan Turing, the brilliant British mathematician who cracked the German Enigma Code during World War II but was driven to suicide because of people's attitudes to his homosexuality. Where the original *Cabaret* glossed over the implicit bisexuality of the Emcee (with Joel Grey appearing as a tuxedo-clad, old-time entertainer) Alan Cumming's performance in the Mendes production was very obviously androgynous. He was far more aggressively decadent, dressed in rolled-up tuxedo pants and jacket, with a bare chest, painted nipples, and a swastika on his buttock that he brazenly displayed in one of the Kit Kat Klub numbers. This overt sexual "deviancy" helped to build the final coup de théâtre. While Nazi homophobia went largely unaddressed in the original production, the final shocking moments of the Mendes production revealed the seductive and flamboyant Emcee as a concentration camp inmate wearing both the yellow star and the pink triangle: on the final drum roll, the show's narrator threw himself on an (imaginary) electric fence (see figure 8.3).

Figure 8.3 Final moments from Sam Mendes's *Cabaret*

Just as changing views of domestic violence colored Hytner's production of *Carousel*, so the harsher tone of Mendes's production can be attributed to a change in public opinion regarding the holocaust itself. Historian James E. Young points out that whereas the instinct in the 1950s and 60s was to find "a redemptory truth, a silver lining at the end of a very long storm" the mood at the end of the twentieth century was very different:

> Now no one wants to find a silver lining. We now believe, in fact, that it is one of the great terrible illusions that we create an automatic redemption in such events as the Nazi era. We are not telling these stories to find a happy ending; we are telling them to find that out of the Holocaust there is no happy ending—there is only a great void.[21]

Interestingly, Mendes's *Cabaret* on Broadway in 1998 coincided with new productions of two other "Nazi" shows that were written within a decade of the musical—*Anne Frank* (1955) and *The Sound of Music* (1959).[22] Both of these new productions took a harsher approach to the era than the originals. The new *Anne Frank*, rewritten by Wendy Kesselman and directed by James Lapine, focused more on Anne's Jewishness and includes a compelling Nazi presence. The original ends with Anne's optimistic declaration that she still believes people are good at heart. The new version concludes with a stark description of her death from typhus at the Bergen-Belsen concentration camp, the details of which emerged after the original play was written. Similarly, where the original *Sound of Music* had shied away from showing swastika armbands (cutting them on the pre-Broadway tryout), Susan H. Schulman's production featured a much more menacing Nazi presence, with swastikas being introduced gradually into the narrative to reflect the insidious takeover of Austria. For the concert at the end of the show, huge flags dropped down from the flies, transforming the theatre space into an intimidating Nazi rally.

Mendes's *Cabaret* received largely enthusiastic reviews in both London and New York. More importantly, there was a general acceptance of Mendes's right to reinterpret the show. In the *New York Times*, arts journalist Michiko Kakutani reflected on the wider significance of these new productions for how we might view the Broadway canon:

> Once in a while, a Broadway revival comes along that is so inventive, so galvanic that theatergoers feel they are experiencing something entirely new. Such is the case of Sam Mendes's daring production of "Cabaret." . . . It is a production, like Nicholas Hytner's 1994 "Carousel," that conclusively demonstrates that

Broadway revivals can be treated like Shakespeare and other classic texts, that in the hands of a gifted director they can be repeatedly reimagined and made to yield new truths.[23]

MATTHEW WARCHUS'S *FOLLIES*

Where the revivals of *Carousel, Oklahoma!,* and *Cabaret* were to a great extent driven by a more contemporary relationship to the themes of the originals, Matthew Warchus's *Follies* (2001) confronted the ghost of an iconic original production and raised important questions about the sense of ownership that surrounds original productions of the musical theatre classics. Despite a relatively short original run, mixed reviews, and audience members walking out every night, *Follies* has acquired a fiercely loyal following. The divergent reactions can perhaps be explained by the coexistent and, at times, competing impulses of the show as a celebration of the Broadway musical and as a social commentary on post-Vietnam America. However, as previously discussed, the central impetus for the show was as a social commentary and Sondheim has explained that the show's setting was intended as a metaphor:

> The *[Ziegfeld] Follies* represented a state of mind in America between the two world wars. Up until 1945, America was the good guy, everything was idealistic and hopeful and America was going to lead the world. Now you see the country is a riot of national guilt, the dream has collapsed, everything has turned to rubble underfoot, and *that's* what the show was about also—the collapse of the dream.[24]

There is no question that this bleaker aspect of the show was present in the original production to anyone who was prepared to look beyond the spectacle. Jack Viertel, who saw the original production many times, recalls that

> the show was almost completely joyless. Even things like the trio of songs "Broadway Baby," "Paris" and "Rain on the Roof"—each one of those deals with a different kind of American dream that's gone sour. I watched audiences walking out—people were getting up and storming out from about an hour ten minutes in and they did for the whole run.[25]

However, the prevailing communal memory of the show is as a celebration. This view rests largely on Michael Bennett's spectacular staging of the musical numbers, which featured statuesque Las Vegas showgirls drifting across the set as ghosts. In Craig Zadan's *Sondheim and Co*, he reveals a central disagreement within the creative team of *Follies*. Codirector and choreographer Michael Bennett's interest lay mainly in creating exciting musical numbers, while librettist James Goldman, director Hal Prince, and composer-lyricist Stephen Sondheim saw the piece primarily as a serious musical drama, using the glamor and spectacle of the flashback numbers to provide a contrast to the characters' present day lives.[26] The affection that many people feel for the show rests largely on the staging and on Stephen Sondheim's pastiche score of show tunes, lovingly preserved in a concert recording that features Broadway veterans such as Elaine Stritch and Barbara Cook. The score has also found a life on the cabaret circuit where songs written as bittersweet character studies take on less complicated meanings: out of context, "In Buddy's Eyes" becomes a romantic ballad rather than a smokescreen for an unhappily married woman and "I'm Still Here" becomes a triumphant song of survival rather than the jaded reflections of a woman who has lived through too much. This upbeat reading of the show was reinforced by the 1987 London production in which the show was substantially rewritten to create a more optimistic story, prompting theatre critic Mark Steyn to complain about the "Jerry Hermanization" of the show and the fact that "what was intended as a cool dissection of nostalgia is, frankly, a great big wallow in it."[27]

The desire to remember the more flamboyant aspects of the original production might help to explain the rather sour tone of some of the reviews of Warchus's 2001 production. Clive Barnes warned readers of the *New York Post* that "this latest folly of a *Follies* ain't optimistic and it ain't splashy. . . . At times it's not only downright dowdy but serious to the point of gloom."[28] There is a similar note of reproach in *New York Times* critic Ben Brantley's comment that a show which is "widely remembered as a ravishing musical elegy for an era in American show business has resurfaced as a small, bleak and pedestrian tale of two unhappy marriages."[29] There was certainly a love of American musical theatre in the pastiche songs, staging, and performances of the original production. But while Barnes and Brantley may choose to remember the show for its optimistic, splashy, or ravishing elements, this is not primarily what the show was about for the creators.

As with any production, the directorial choices will result in emphasizing some aspects of a show more than others. However, far from being a perverse distortion of *Follies*, Warchus's production was in many ways a faithful rendition of the show as it was written rather than how it is now remembered. Instead of emphasizing the nostalgic elements of the show,

Warchus based his interpretation on its darker undertone as manifested in the published libretto and score and on early drafts, which he describes as Chekhovian in mood. He recalls one draft in particular in which Sally is pregnant by Ben when he leaves her for Phyllis, and in which Ben and Buddy engage in a sword fight until Sally comes in with a gun. "It's like *Uncle Vanya* or something because I think she shoots and misses—she tries to shoot Buddy."[30] Rather than being a love letter to the theatre, Warchus's interpretation of *Follies* was as an enormously ambitious musical tragedy about "the slide from optimism and innocence to decay" more reminiscent of *King Lear* than musical comedy:

> It's also about being on that journey and how that journey is such a fast-moving one, and it's very easy to make a lot of mistakes before you've had a chance to think about it. And the road you're taking is informed by the number of mistakes you make, not realizing things in time—by not having knowledge, or an overview, about life until it's too late. So you get wiser after the event: you're losing your life but you're gaining wisdom. [31]

Embracing the show's tragic dimensions is not without its challenges and the Chekhovian analogy raises issues of tone and energy. Warchus's decision to cast the four leading roles with actors who had little or no singing experience also presented stylistic problems, making it difficult for the show to soar emotionally at strategic moments. But his concept was rooted firmly in the text and was no more a distortion than the versions that overemphasize the more celebratory elements. While sacrificing some of the emotional release, Warchus achieved a more psychologically motivated interpretation of many scenes, allowing the characters' mood to dictate the staging rather than aiming for an impressive visual effect. In the Zadan interviews, Bennett recalled the opening number in the 1971 production as a visual feast: "When that curtain went up it was the most exciting thing ever . . . the most fabulous set, the most beautiful costumes, the most gorgeous chorus girls, the most brilliant lighting . . . it was just incredible!"[32] By contrast, the 2001 audience entered a Belasco Theatre hung with ghostly gray drapes and the lights went up on a bare stage with only a raised scaffolding platform along the back wall. A lone actor with a flashlight shone a narrow beam of light around the stage and into the audience while vague echoes of noises and voices from the past faded in and out.

Warchus's psychological approach was also evident in the staging of "The Right Girl," Buddy's solo number about the bitter irony of having both a wife and a mistress and loving the one who does not love him back.

Bennett's approach was to use spectacular dancing for the number. Ted Chapin was an assistant on the original production and in *Everything Was Possible: The Birth of the Musical Follies* he talks about the original Buddy, Gene Nelson, who was formerly one of the best tap dancers in America. When he was unable to stay on beat, Bennett "came up with new ideas, including additional swings around poles and an astonishing leap from one high platform onstage right to another. He was also determined to devise an ending that would both act as a choreographic punctuation and get a huge hand."[33] By contrast, Treat Williams's performance in Warchus's production (choreographed by Kathleen Marshall) was psychologically motivated and derived its impact from the unpredictable energy of a man in the grip of an overpowering emotion. While much of the song was sung out toward the audience, it also included moments of interaction with his young self (toward whom he expressed contempt for being naïve), conversations with his mistress, a sudden outbreak of violence (knocking over a chair), and a moment of total breakdown out of the spotlight. The number ended as he threw a glass against the wall in frustration and anger, at which point Sally appeared and the scene continued before the audience had a chance to break the mood with applause. For the audience, it was the unsettling experience of watching somebody having a breakdown rather than the thrilling experience of watching a star performer executing inventive choreography.

DAVID LEVEAUX'S *NINE*

Like Warchus's *Follies*, David Leveaux's *Nine* offered a radically different approach to the show than the original production. Based on Fellini's film *8 ½*, *Nine* centers around celebrated director Guido Contini who is blocked on a movie and is feeling the strain of juggling complicated relationships in his professional and personal life. It goes on to illuminate how Guido is being pulled in different directions and to explain how he got to his present state through his interaction with the figure of his younger self and with the central women in his life, including his mother, his wife, his past lover and muse, his present lover, and a prostitute who initiated him into sexual awareness. Through these interactions (part real time, part memories) he is brought, finally, to a crisis point where he must let go of some people in order to move forward. The story is not told chronologically but piecemeal, with real "present day" events triggering either Guido's memories or other characters' sung reflections on their relationships with him. Musical numbers fulfill several functions (as memories, inner reflections and—in the second act—a film production number) while dialogue is almost exclusively used for present-day scenes.

The original staging of *Nine*, like that of *Follies*, had become inextricably associated with the show itself. The fact that it premiered as late as 1982 ensured that Tommy Tune's high-concept black-and-white production— with all the characters based on plinths scattered around the stage—was still fresh in the collective memory. However, where the original *Follies* was an attempt to realize its complicated dual themes, Tune's original staging of *Nine* was based on showmanship rather than on a real understanding of the libretto and score. In his autobiography, *Footnotes*, Tune admits his lack of connection to the show other than as a collection of terrific musical numbers, confessing that "I was always aware of the smoke and mirrors that I'd created to divert the viewer from the fact that it lacked a plot."[34] Leveaux recalls a startling moment at the opening night of his production when Tune turned to him and admitted frankly: "This is the first time this has ever made any sense to me."[35]

Tune's lack of connection to the material is an illuminating reflection of the difference between visually based musical comedy and thematically driven musical drama. Tune's approach to musicals was that of a choreographer-director in the tradition of Bennett and Fosse, with a focus on overall visual effects and stand-alone choreography that prompted Frank Rich to remark that the real emotion of the original production was Tommy Tune's love affair with the theatre.[36] But while *Nine* is set in the world of show business, it is the esoteric film world of 1960s Italy rather than the brash, upbeat world of Broadway musicals and there is a European sensibility in the show that is at odds with the traditions of Broadway musical comedy. Leveaux's production was, by contrast, a more thematic response. He is resistant to the labeling of British directors as "book driven" in the sense of the libretto without the score; before *Nine*, his background was not only as a director of plays but also of opera and the lyrical and nonliteral possibilities of theatre are integral to his approach. He cites as early role models both "high art" names such as Merce Cunningham and Pina Bausch and that of Michael Bennett, emphasizing their ability to manipulate the nonverbal elements of a production:

> That's one of the fascinating things about how the theatre works—how the stage works—because sometimes you are operating at this purely intuitive level where someone doesn't understand why they cried. That's also why I love the musical as a form. It's not to do with getting solemn about text; it's to do with responding

to what I would call the deep gesture of music, which is that music operates at all sorts of levels which are inherently dramatic if it's dramatic music.[37]

Leveaux's *Nine* demonstrated his dual background in opera and drama, allowing his interpretation to emerge from a response to both libretto and score. His staging can largely be traced back to his identification of two central themes. The first is the idea that "there's a dialectic going on here which is about longings, which is about something lost that's not entirely recoverable."[38] Tune's original 1981 production was contemporary for the time with a sleek, austere, black-and-white designer look. Leveaux's decision to set his production in the 1960s, the era of Fellini's movie, was a move calculated to make sense of the show's tone and subject matter for a contemporary audience:

> I put it down to innocence. I don't mean naiveté, but innocence. The whole thing about the 1960s was that things were possible and the core of *Nine* is innocence . . . there is this kind of optimism and celebration of all kinds of different forms of love some of which are admittedly probably not sustainable ultimately, but unless you've got that kind of glint of a world where sex and love are good things, as in the 1960s, then this is hard to express . . ."[39]

In the opening number of the show, where Tune had Guido conducting the women (an image of control in a musical about a breakdown), Leveaux centered the action around a large dining table to represent Guido's attempts to gather together the central people in his life:

> What's happening in the first few moments of *Nine* is that he's looking for something. He doesn't yet know that he's fallen off a very, very high building. He's looking for something and in so doing he's checking back through his life. Each one of these people may have the answer to the thing that he's somehow missing. So I thought all those women would come to a place in his mind in which all those differences could be held together in a single gesture, which is a dinner table. Plus it's communal, plus the idea of coming together at a dinner table is fundamentally celebratory and is to do with relationships.[40]

Equally, when the women in Guido's life descended a spiral staircase at the beginning and ascended at the end it was an attempt to highlight one of the central themes: "It's ultimately so we feel they're ghosts—these girls coming down that spiral staircase are like ghosts and instantly you see them it should open up a ball of nostalgia and loss in some way."[41]

The second theme that stimulated Leveaux's production was that of Guido's relationship with women. Where Tune saw a lack of plot, Leveaux saw a celebratory fantasia on the theme of love, bookended by the story of a marriage.

> *Nine* is not a facile story about how at the end of the day you have to grow up and that those fleeting romantic relationships are superficial in comparison with the real one. The reason Guido Contini is a great director is because he loves women in a certain way and is able to film them in a way that reveals something—often quite radically. There is a genuine vitality and continually elusive light in those relationships he has with women, which are sometimes 30 seconds long."[42]

In Leveaux's eyes, the women in Guido's life are intelligent, independent women who do not need him but who choose to want to be with him. While the female characters in Tune's production were largely one-dimensional (the wife, the mother, the lover), Leveaux's production offered a more complicated psychological exploration of the relationships through individual performances and, particularly, their spatial relationships. Guido's song "Only with You" starts and ends as a testament of his love for Luisa but expands to refer to his other two loves. In the Tune production, Guido simply walked from Luisa to his mistress (sitting on her cube) and back during the song. In the Leveaux version, Guido was arranged holding both his wife and mistress with Claudia as a third presence without any of the women seemingly aware of the others—a piece of staging that succinctly expressed Guido's complicated relationship to love and to the three women.

The difference between the two directors' approaches was also apparent in their staging of the show's most memorable musical number, "Call From the Vatican," in which Guido pretends to take a call from the Vatican that turns out to be a seductive call for attention from his lover Carla. In the Tune version, Carla (Anita Morris, an accomplished gymnast) did stunning acrobatics in a lacy body stocking, using the cube as a prop in an inventive display of physical dexterity. The number was performed in a solo spot with Guido watching from the sidelines. In the Leveaux version, Carla (Jane Krakowski) was lowered from the ceiling wrapped in a long white sheet until she reached the table where Guido was sitting with the telephone and writhed suggestively around the table, wrapping herself in the phone cable and caressing his face with it before bidding him farewell and exiting into the flies again as the call ended—staging that emphasized the sensuality of their relationship but also her loneliness in a hotel bedroom.

Even more telling, perhaps, was the depiction of Luisa in the moments following "Call from the Vatican." In the Tune version, Luisa registered no particular response to the call; not only did this cast her as rather stupid but it also made the audience a collaborator in Guido's secret. In the Leveaux production, Guido grinned sheepishly into the phone and feigned innocence, completely unaware that his wife knew exactly who was on the other end of the phone line. This changed Luisa from dupe to a mature adult, and conveyed the sense of long-standing resignation with which Luisa dealt with the situation, thus explaining not only how she had coped with him for so long but also why she finally left him. It was a clear example of a directorial approach that emphasized underlying theme and psychology rather than musical comedy staging.

The approaches examined in this chapter are not universally applicable and I am not suggesting that all musicals would lend themselves to this kind of reinterpretation. In particular, some musical comedies have specific performance traditions so deeply embedded in the tone and the structure of the shows themselves that it can be difficult to apply new kinds of staging without unraveling the show itself. Equally, there is a place for reproducing and celebrating the work of past choreographers and directors, for example, through tribute shows (as with *Jerome Robbins' Broadway* or *Fosse*) or by reviving specific productions, as with opera and ballet. However, it seems to me that if the musical canon is to continue to have relevance and meaning for contemporary audiences then directors must be allowed the freedom to explore new interpretations of the classics and to find different ways of releasing the themes and ideas that are embedded within the published texts. The new directorial approaches discussed in this chapter have so far been mainly applied to musicals from the 1940s, 50s, 60s, and 70s. However, it will hopefully also help to influence the way in which we start to contemplate the afterlife of musicals from the 1980s and 90s, allowing us to continue engaging with the canon as living texts rather than fossilized reminders of how the musical used to be.

9. The Legacy of the 1980s and 90s ✥

Whatever our individual tastes, few theatre historians would deny that the 1980s and 90s have been decades of great change for the musical. The exact legacy of this era and what it means for the future of the art form still remains to be seen, but there have already been significant changes in how we approach the creation of new musicals in Britain and America. In terms of musical drama, it seems to me that there are four areas in particular that will prove highly influential in how the art form develops in the coming years: the increased crossover between the musical and other performing arts; changes in commercial producing; the growing relationship between nonprofit and commercial producers; and finally the question of where the next generation of musical theatre directors will come from.

THE MUSICAL AND OTHER PERFORMING ARTS

At the start of the twenty-first century, the Broadway community experienced a palpable sense of relief and triumph as old-fashioned American musical comedy made a comeback starting with *The Producers* (2001) and followed by *Thoroughly Modern Millie* (2002), *Hairspray* (2002), and *Dirty Rotten Scoundrels* (2005). While these shows vary in tone, style, and source material, they all feature the flamboyant characters, star performances, upbeat show tunes, and feel-good storylines of musical comedy and represent a return to book musicals that emphasize witty dialogue and clearly defined songs rather than the through-composed format. In his book *Ever After: The Last Years of Musical Theatre and Beyond*, Barry Singer succinctly captures the feeling of nostalgic relief that washed over Times Square:

> *The Producers* arrived on Broadway in April 2001 and gleefully marched the Broadway musical backward in time. The schtick, the girls (both young and old),

the tunes, the governing taste, all were determinedly old-school and audiences loved it. So did critics. As the screenwriter and entertainment industry pundit, William Goldman, put it in *Variety*: "Come back to the theater. All is forgiven."[1]

However, while these shows represent a return to 1950s and 60s musical comedy, Broadway itself has changed considerably and these shows coexist with musicals that have a very different sensibility. Broadway today is increasingly a receiving house not only for productions but also for artists who originated Off Broadway, in the nonprofits and abroad. Since the year 2000, musicals on Broadway have included *Urinetown*, the unlikely Fringe Festival hit that transferred from Off Broadway; *Avenue Q*, a social satire using puppets in the style of *Sesame Street* that transferred from the nonprofit Vineyard Theatre; *Caroline, or Change*, a musical drama transfer from the Public Theater that has the lead character interacting with actors representing a washing machine, a radio, a bus, and the moon; *Bombay Dreams*, an import from London featuring a Bollywood score and choreography; *La Boheme*, based on Baz Luhrman's 1990 Sydney Opera House production; *Movin' Out*, a dance show featuring a pianist-singer that was conceived and staged by modern dance choreographer Twyla Tharp; and Japanese director Amon Miyamoto's production of *Pacific Overtures* using kabuki traditions.[2]

With the influx of these different artists and works, there has been a gradual acknowledgment that while the developments on Broadway in the 1940s, 50s, and 60s constituted an enormous leap forward for the musical, they do not define its parameters today. In 2002, *New York Times* journalist Don Shewey wrote an article entitled "Just What Is a Musical? Broadway Has a New Definition" in which he depicted the Broadway debuts of Baz Luhrman and Twyla Tharp as representative of a larger cultural shift. While Broadway shows such as the *The Producers* and *Thoroughly Modern Millie* were created by "veteran insiders at the peak of their form or newcomers following in their footsteps," he argues that Tharp and Luhrmann offer an alternative approach to the musical:

> With their idiosyncratic artistic visions, they are aiming for an audience that does not necessarily see every Broadway musical. Exuding a recklessly experimental rock'n'roll energy, they hope to lure those who might otherwise never suspect that ballet or opera could appeal. And they both see Broadway theater as a kind of big, friendly carnival tent where audiences for different art forms can cross-pollinate.[3]

CHANGES IN COMMERCIAL PRODUCING

Within the commercial sector, there have been significant shifts since the start of the 1980s. Marketing of shows is perhaps the most visible change, and in this respect Cameron Macintosh has left a lasting legacy with his emphasis on merchandising and his radical approach to poster design, using a single identifiable logo and ensuring that only two names appear on the poster: the show's and his own. The latter innovation reflects a fundamental shift away from the star vehicle to shows in which the production itself is the main attraction. This is particularly important given the escalating costs of commercial producing, with many Broadway musicals now costing $10 million or more. While it can take years to recoup the cost of a musical, most of today's biggest stars balance theatre work with the far more lucrative worlds of film and/or television and cannot afford to sign up for more than a year at the most. The distance traveled in this respect was clearly apparent in March 2002 when the stars of *The Producers*, Nathan Lane and Matthew Broderick, came to the end of their contracts. Unlike shows such as *Cats*, *Phantom of the Opera*, or *Les Misérables*, the producers of this show had reverted to the musical comedy tradition of emphasizing the stars as the chief attraction, featuring their names and images prominently on the publicity materials and promoting them as a must-see double act. This presented a problem when they left, and a reconfigured poster started to promote the show on the basis of the production rather than the star performers.

With the rising costs of producing a new show on Broadway and in the West End, commercial producers have had to find new strategies for presenting work in a financially viable way. One solution has been to build shows around the back catalogue of pop music in the expectation that people will come to hear what they already know—a twist on the well-established practice of prereleasing a show's hit songs, a common practice in the Golden Age and all the way up to *Evita*. This trend has been seen as an unhealthy one and a sign of "dumbing down" the musical. I would argue that there is a place for good examples of this alongside other kinds of shows—*Mamma Mia!*, which helped to launch this trend, strikes me as a very witty and original piece of work unlike some of the less imaginative shows that followed. However, while the book of the show may be new, the music is not and the main problem with this trend is the fact that it is shutting out the next generation of composers and lyricists from the commercial theatre.

Another solution for commercial producers has been to scale down their productions either in the hope of making a profit through an open-ended Off-Broadway run or in order to test the waters before plunging into the economic minefield of Broadway and the West End. This has led to the growth of commercial Off-Broadway and fringe musicals. In New York, examples include *Urinetown*, *Batboy the Musical*, and *The Last Five Years*—all interesting, original, and risky shows that could not easily have opened in large commercial houses. *Urinetown* is a particularly interesting case in point. A surprise hit of the New York Fringe Festival, this original, irreverent satire of both musical theatre and Brechtian drama traded on informality and grunginess and would not have been well served by being placed in a glamorous Broadway house. Instead, the producers built up audiences through an Off-Broadway run before transferring it to the Henry Miller, a particularly rundown Broadway theatre that the producers were able to distress even further to create the right audience experience.

Both London and New York have also benefited from the establishment of musical theatre festivals designed to showcase new work at affordable costs. The New York Musical Theatre Festival (NYMF) was launched in September 2005. As a founding producer of the festival my outlook is not completely objective, but it seems indisputable that NYMF has quickly become a significant part of the musical theatre ecosystem by providing a stimulating annual meeting ground for the musical theatre industry and a high-profile launch pad for new musicals of all descriptions. One of the first beneficiaries was *Altar Boyz*, a musical built around a spoof Christian boy band, which was produced on a modest budget in the first festival, proved a runaway hit with audiences, and subsequently transferred to a long open-ended run at the commercial Off-Broadway Dodger Stages. In Britain, there has not been a new musicals festival on the same scale, but rather a number of smaller outlets. In South London, the Greenwich Theatre runs an annual Musical Futures festival offering a showcase of new musicals and in 2002, 2005, and 2006, the Cardiff International Musical Theatre Festival showcased new work through its Global Search for New Musicals. September 2006 saw the launch of Perfect Pitch, a London-based musical theatre festival based at the small Gatehouse Theatre in North London and primarily intended as a way to connect commercial producers with new musical theatre projects at all stages of development.

THE NONPROFIT/COMMERCIAL
RELATIONSHIP

Of all the shifts in developing and producing new musicals, the most significant is unquestionably the changing relationship between the commercial and nonprofit sectors in America. The theatrical landscape has changed drastically since 1984 when Playwrights Horizons was chastised for inviting Stephen Sondheim to work in a nonprofit theatre. Today, organizations such as the National Alliance for Musical Theatre can attest to the changed infrastructure, offering networking opportunities, conferences, and an annual festival of new musicals for its 142 member organizations, comprising nonprofit theatres from across the United States and abroad.[4] In addition to producing venues such as the Goodspeed Opera House, which are entirely dedicated to musical theatre, there is a large network of regional nonprofit theatres that is committed to creating and/or programming musicals as part of their ongoing artistic mission, either through their own development programs, in partnership with commercial producers or as coproductions with other nonprofit theatres. Funding for nonprofit musical programs has enabled more experimental work to continue, while projects that have more populist appeal can be financially enhanced by commercial producers. The result has been a broad range of interesting work, including *Caroline, or Change*, which was developed and produced at the Public Theater before transferring to Broadway; *Avenue Q*, a much more commercial but very quirky musical, which built up a following at the nonprofit Vineyard Theatre before transferring to Broadway (where it was the surprise winner of the Tony Award for Best Musical) and subsequently Las Vegas and the West End; and *The Light in the Piazza*, a lyrical and musically sophisticated exploration of love, which was first staged in nonprofit theatres in Seattle and Chicago before arriving at the Vivian Beaumont Theatre in New York. The Beaumont is part of the nonprofit Lincoln Center Theatre but counts as Broadway because of its size and is, therefore, eligible for Broadway's Tony awards. *The Light in the Piazza*, a musical whose tone and subject matter is not obviously commercial and would have struggled to open on Broadway, was acclaimed with six major Tony awards—for best score, orchestrations, leading actress, set, costume, and lighting.

There are, of course, some potential problems with the increasingly symbiotic relationship between commercial producers and the nonprofit sector. As Zelda Fichandler pointed out in an article for *American Theatre*, "It is a good idea for us to remember that while to experiment means to test or to try, to try is not at all the same thing as to try out. Our theatres are destinations in themselves, crucibles in which to test a vision of the world that is ours, not someone else's."[5] The reverse is also true: the reliance on nonprofit theatres for developmental work or initial productions can result in material and in staging that is not ideally geared toward the Broadway theatres. Jack Viertel, who has worked in both sectors, points out that there is an insurmountable physical hurdle:

> You can't do in any resident theatre what you can do at the highest level on
> Broadway in terms of scenery, costumes, scenic transitions and special effects. It's
> not just a matter of money—the physical limitations of resident theatre build-
> ings and shops won't allow it. The buildings weren't designed to accommodate
> the mission of doing big Broadway musicals with mechanization and fly galleries
> and so forth. So the shows that have been developed in resident theatres by and
> large have been physically modest because, having conceived them for a theatre
> that has limitations, it's not worth reconceiving them for Broadway—you simply
> dress them up a little bit more.[6]

In America, the growth of nonprofit musical theatre development programs has been the result of external funding from a variety of sources,
including subsidies from state-level funding bodies, charitable foundations,
and tax-deductible schemes that encourage gifts from individual donors. In
Britain, arts subsidy is still more centralized through the Arts Council where
funding for the development of new musical theatre (as opposed to the
more operatic music theatre) remains largely nonexistent. As a result, the
major regional theatres that might have the ability to play the same role as
the American nonprofits simply do not have the time, money, or the staff to
cope with the special demands of developing and producing new musicals.
This is not to say that there is no activity and some theatres such as the
Royal National Theatre, Greenwich Theatre, the Theatre Royal Stratford
East, and the West Yorkshire Playhouse have all managed to include new
musicals into their work. There are also organizations to support the development of new musical theatre, including Mercury Musical Developments
(for writers) and Musical Theatre Matters (for producers, directors, and
other "creatives"). However, with limited destinations for new work in
British theatres it is a very different situation from that in the United States.
The Arts Council has recently shown signs of recognizing musical theatre as
an art form worthy of funding; however, it remains to be seen whether and
how this will translate into sustained funding and what role it will play in
creating a healthier ecosystem for new musical theatre in Britain.

THE NEW MUSICAL THEATRE
DIRECTORS

The changes outlined earlier have already changed the landscape of musical
theatre. In addition to creating a broader definition of musical theatre and

establishing new producing models, the 1980s and 90s have also changed the idea of musical theatre directors, not least due to the work of the artists discussed in the previous chapters. In Britain, the increased musical theatre activity of the 1980s and 90s has opened the way for more top theatre directors to tackle musical theatre. Since the late 1990s, the directors of musicals in the West End have included opera and classical theatre directors Deborah Warner (*Mama Mia!*), former Artistic Director of the RSC Adrian Noble (*Chitty Chitty Bang Bang*), former Artistic Director of the Royal National Theatre Richard Eyre (*Mary Poppins*), and drama and film director Stephen Daldry (*Billy Elliott*) as well as familiar figures such as Trevor Nunn (*The Woman in White*), Nicholas Hytner (*Jerry Springer—The Opera*), and Matthew Warchus (*Our House, Lord of the Rings*). Some of these shows have continued the traditions of musical drama, notably *Jerry Springer—The Opera* (whose title alone challenges cultural hierarchies), *Billy Elliott*, which deals with the 1980s British miners strike, and *Mary Poppins*, in which Eyre's staging helped to bring out the darker psychological aspects of the well-known story.

On Broadway, producers have now largely accepted that the new American musical theatre directors will not simply be rising through the ranks of Broadway shows in the tradition of Jerome Robbins, Harold Prince, Gower Champion, Michael Bennett, Bob Fosse, Tommy Tune, and Susan Stroman. This is largely due to the shift toward fewer new musicals, smaller casts, and longer runs for hit shows, which means that there is a smaller number of performers to draw from. The problem is compounded by the fact that directing a single Broadway show can now take up several years of a director's professional life, from workshops and tryouts to Broadway run and the subsequent tours.

The initial response of Broadway producers has been to hire experienced British theatre directors such as Nicholas Hytner (*Sweet Smell of Success*), Michael Blakemore (*Kiss Me, Kate*), Sam Mendes (*Gypsy*), Matthew Warchus (*Follies*), and David Leveaux (*Nine, Fiddler on the Roof*) whose extensive work in British subsidized theatre have made them obvious choices for both classical play revivals and musicals. The other response has been to look to the American nonprofit sector where directors can build up their skills and their reputation more quickly. In the last few years, Jack O'Brien and Joe Mantello have become the directors of choice for new musicals although they were both best known for their work on drama in the nonprofit sector. Since 2000, O'Brien (the artistic director of the non-profit Old Globe Theatre in San Diego) has balanced his musical theatre work (*The Full Monty, Hairspray, Dirty Rotten Scoundrels*) with drama

(*Imaginary Friends*, *The Oldest Living Confederate Widow Tells All*, *The Invention of Love*, *Henry IV*, and *The Coast of Utopia*). Joe Mantello directed two very different musicals in the 2003–2004 Broadway season: the commercial *Wicked* and a revival of Sondheim/Weidman's *Assassins* through the Roundabout Theatre Company. Prior to that his experience had been as an actor (including the Broadway premiere of *Angels in America*) and more recently as a play director in the nonprofit sector (*Love! Valour! Compassion!*, *Take Me Out*) and on Broadway (*Design for Living*, *Frankie and Johnny in the Clair de Lune*). The most prominent example of how this might affect the musical is arguably *The Lion King* as directed by avant-garde theatre director Julie Taymor. Rather than simply recreating the Disney cartoon (as with *Beauty and the Beast*) or turning it into a traditional musical comedy, Taymor created a whole new theatrical vocabulary for the story, representing the animals through sophisticated puppetry and adding new, more authentically African music to the Elton John score.

Clearly there are repercussions to this shift. On the one hand, it opens up the possibility of new directions for the Broadway musical, both in terms of dramaturgical approach and in terms of the development and rehearsal process. In particular, there has been a change in the old-fashioned model of the out-of-town tryout, which preceded the Broadway opening. Jack Viertel points out:

> There are no longer a lot of directors who know how to work in the old style. Jack O'Brien did it on *Hairspray* and Susan Stroman did it on *The Producers* but it's done less and less. It was very interesting watching Tommy Tune do *Grand Hotel* in Boston as one of the last people to come from the generation of directors who knows how to fix a show on the road: rehearsing one version in the afternoon and playing another version at night, and then teching the new version the next afternoon and playing it that night. Directors from the non-profit world often have no idea how to do that—and why would they want to try it? Their notion is that you do the show at a resident theatre, you look at it, you close it, you fix it up on paper and then you put it up again. It's not necessarily a negative evolution, although I do feel that old-fashioned showbiz is perpetually threatened by these kinds of developments.[7]

Instead of the out-of-town tryout, new musicals are now more commonly developed through readings and workshops and will often have their initial production in a nonprofit theatre, either through the theatre's own musical theatre program or through an enhancement deal between the theatre and a commercial producer.

Whatever our musical and theatrical tastes, it seems clear that the 1980s and 90s represent a vital chapter in the ongoing evolution of musical theatre as part of the cultural landscape. As we start to come to grips with these changes and to assess their long-term impact on the art form, it seems to me that we cannot simply sideline or dismiss anything that does not fit into the traditional historical narrative of the Broadway musical. I would argue that the single most important shift of the 1980s and 90s is not of size (whether in terms of box office, scenery, or marketing campaigns) but of artistic breadth. The great legacy of the artists and works discussed in the previous chapters is the increased dialogue between musical theatre, drama, and opera and between commercial, subsidized, and nonprofit theatre. In order to engage critically with this work we cannot simply maintain the definition of musical theatre established in the Golden Age of the Broadway Musical. The ongoing development and recognition of musical drama will depend on whether writers, designers, and directors are encouraged to continue interrogating the art form; on whether producers and funders are able to support such work; and on whether critics and historians are prepared to move away from restrictive terminology and toward a more inclusive analysis of musical theatre in the 1980s, 90s, and beyond.

Interviews ✧

B etween January 2003 and May 2004, as part of my research for this project, I interviewed a wide range of theatre professionals in New York and London, including writers, directors, producers, arts funders, and other people involved in the creation and production of new musicals in the 1980s, 90s, and beyond. My aim was to build up a full picture of this era in musical theatre history from people who were directly involved and to include some of their insights in my study. As I progressed, I realized how uniquely privileged I was to have had the opportunity to travel between these two musical theatre capitals and to have been granted access to such a diverse and distinguished group of musical theatre professionals.

Rather than simply providing a backdrop to my analysis, I have come to feel that these interviews constitute a historical record in their own right and that they might be a valuable resource to other musical theatre artists, historians, students, and audiences. In the following pages, I have, therefore, included transcripts from a representative sample of the interviews that I conducted. They are all focused around the central questions of this book—musical dramas and the role of drama directors in developing them—but within these topics they cover a wide range of issues. The interviews have been reproduced in the order in which they occurred, allowing the reader to follow some of my developing lines of enquiry. I have also noted the date and place of each interview as they are essentially a snapshot of a particular moment in theatre history and in the professional lives of my interviewees.

It was an enormous pleasure for me to compile these interviews. By sharing the transcripts with a wider readership it is my hope that they might give similar pleasure to others and help to foster a greater interest and engagement with the rich and varied legacy of British and American musicals in the 1980s and 90s.

INTERVIEW WITH ADAM GUETTEL
(April 9, 2003, New York City)

Adam Guettel is a composer-lyricist and has been widely hailed as a pioneer of the new American Musical. My interview focused on his song cycle Myths and Hymns *(initially staged as* Saturn Returns*) and* Floyd Collins. *At the time of this interview he was about to start the first previews of* Light in the Piazza, *a show that subsequently had successful nonprofit runs in Seattle and Chicago and went on to win six major TONY Awards when it transferred to New York, including best score and best orchestrations.*

What do you look for in a collaborator? Is there a reason that you have ended up working with writer-directors?
My guide for choosing a collaborator has to do with his or her sensibility and aesthetic and our shared aspirations in terms of music drama. When I write I have to think about how something might work on stage or how it might be directed and I am drawn to people who work that way.

I am very interested in the question of authorship in the musical and how this intersects with the collaborative nature of making musical theatre. With some classic musicals, the original staging is sometimes contractually regarded as part of the show for licensing purposes and there seems to be a recent shift toward granting royalties to a number people other than that book writer, lyricist, and composer. What is your perspective on this?
Since Hal Prince and Michael Bennett and Bob Fosse the idea of the director as author has become more accepted. Sometimes it's legit. A true originating concept is authorial in nature and requires time and energy and brings responsibility. But now there's an unhealthy trend in the theatre whereby directors, choreographers, lighting designers, costume designers, and musical directors want a royalty in perpetuity based on the unfair premise that because they contribute ideas during three or six weeks of rehearsal and preproduction they are somehow authors. A director or designer's contribution is of course terribly important, but essentially interpretative.

An author spends years alone inventing from nothing—every word, every note is a decision. The author creates the world of the piece—the tone, the structure, the weight, the rhythm of tension, and release—and the author bears by far the greatest risk to his reputation and has spent by far the most time. All the basic elements of drama are created by the author. They are then interpreted. Directors and designers inflect the material. What is taking place is collaboration. To monetize this beautiful process is to discourage it.

When you are writing a piece, at what point do you start thinking about the physical production in terms of scale and possible venues?
The economics of producing theatre in this country are such that you have to start thinking about how it's going to be produced and what your forces are going to be from very early on. In the case of *The Light in the Piazza* I knew from the start that it was going to be a small cast (eight people) and a small instrumentation (four players). We would be thrilled but very lucky to have a successful commercial run in New York. We're more likely with this piece to have a life in the regionals and the colleges and conservatories and for that reason I want it to be very transportable and very economically feasible. I was thinking about that from the beginning.

For *Floyd Collins*, how far were you influenced by the film?
I never saw the movie. I haven't seen the film for *Light in the Piazza*.

Tina Landau has said that for her *Floyd Collins* is about what it means to be American now. Does that go for you too?
That's part of it. I think it does resonate.

What attracted you to the *Floyd Collins* story?
On some sort of subconscious level I think I was attracted to this story because I spend a lot of time in my own metaphorical hole. . . . Also (I don't know that I was aware of this when I was involved with the project.) I think I wanted to work on a story where someone had aspirations for something beyond their reach, failed at attaining those goals, and still had a noble life. I think I wanted to portray that sequence, that trajectory. Because (again without knowing it) it sort of relates to what I'm doing in the theatre given what my background is.

You mean your grandfather, Richard Rodgers?
Yes. And I think on a more dramatic level I was attracted to the opportunity to derive a score from a certain kind of sound—a certain time and a certain place. It was also (for the music below ground where Floyd is) the opportunity to do whatever I wanted and make that cave sound however I wanted. It seemed sort of free of association whereas above ground it was going to be fairly connected—loaded with association—which was another challenge.

Could you talk about your collaboration with Tina Landau? I gather that you started with a paragraph in a book describing the Floyd story. How far did you work on structure and themes together?
Every time we met we would talk about structure and the importance of those things. Some days we would meet and the societal strands versus family,

personal, spiritual strands wouldn't come up; some days it was sheer structure working out how and what will happen in this particular scene and spotting the songs. We did everything together and she wrote the spoken words and I wrote the sung words. But in terms of weighing out for each of us what our homework was, what our assignments were, we would do that together and the tone of the piece would be developed during those conversations. And I would say that much less with Tina than with Craig did we discuss directorial ideas while we were writing. We were more concerned with telling the story cleanly and just sort of stayed in a writerly place, me and Tina. And then when she took over as a director she put an absolutely beautiful interpretation on this story. But I think that there was a kind of demarcation there. She may without saying it have been thinking directorially.

There was such a beautiful simplicity in the staging.
Tina comes from a kind of nonrealist tradition. Also, the more we leave to people's imagination the more opportunity we give them to insert exactly what they want to see.

Was the writing affected by the original stage at American Music Theatre (now the Prince Music Theater) in Philadelphia?
Somehow we had to think in terms of the reality of the stage but most of the time we would allow ourselves to be connected to the emotional through line and worry about the practical matters when we got to them.

I was struck by the fact that *Floyd Collins* was small-scale, using simple devices like lighting shadows and a small cast to tell what could have been a big story.
It was part economics, part aesthetic decision. One thing can stand for many and one kind of behavior can be symbolic. Those efficiencies can be satisfying. They can be found in many pieces of theatre, of course, even when resources are not limited.

***Floyd Collins* followed an interesting path, from the American Music Theatre in 1994, Playwrights Horizons in 1996 to the Old Globe, Prince Musical Theatre, and the Goodman in Chicago. Was there an overall tendency in the changes that you made?**
The audience in Philly made us aware that the opening of the story didn't provide a road map, a sense of what's to come. It did not give us a sense of what this central eponymous character wanted. It was stream of consciousness, associative. The major change from the AMTF to Playwrights Horizons was to conventionalize the opening: it was less experimental, it was shorter and catchier and more direct about what Floyd wanted.

What happened in rehearsal for Chicago was that we just felt the second act was a little overdone and not generous enough to the audience. It was almost too claustrophobic and we didn't give them any space to consider what they'd seen and what they sensed was going to happen. It didn't give them a moment of respite so we restructured the second act so the "Ballad of Floyd Collins" is sung in its entirety. It's a moment of reflection, really, which the show kind of needs to get some air in there. And "Tween a Rock and a Hard Place" in the first act was replaced. It was the only song I ever wrote where I did the lyrics first. I haven't done it since. It created a very square kind of music—it was the most stereotypical music and the title was a cliché.

The goal of that song was to do something that's a reasonable thing to try, to show with locals the kind of thing that would happen later in the show with reporters. It was a song where these three locals say: "Yeah, Floyd's in trouble, that's for sure, but I've been in much worse . . ." And they proceed to trump up their lives. I wanted to show that at a local level, so that when we got to the reporters who trump up the facts we wouldn't just compartmentalize and vilify them in a generalizing and very pre-dictable way. Everyone in the show, including Floyd, has a poetic and a mercenary side (except perhaps Nellie: she's just pure). We see in Homer a tenderness and a great loyalty to his brother that eventually breaks: he can't deal any more, he goes into the movies. Floyd himself is trying to make it big but he has a poetic window into that like when he's describing the cave: "layerin' on slow / Like a stone history book/ With a couple of sentences about me." He wants to find something beautiful. He wants to transcend his circumstances and make a pot of money off of nature. He's mercenary and poetic. It's meant to plant a seed—that the reporters aren't the bad guys and the locals aren't the good guys. "'Tween a Rock and a Hard Place" did not do that. What the audience got was: "We've just spent 20 minutes with one guy and now who the hell are they? We don't know who they are but we think they're going to be really important to the story." But we hardly hear from them again. So the song was a cliché—both musically and lyrically—and it didn't do enough to set up that world economically and socially. The new song "Where a Man Belongs" [written in 1999] does it better, I hope.

How do you feel about having different productions of the piece?
I like these shows to be done and reinterpreted. That's where the separation of author and director really comes in handy and breathes life into a piece. Why should it be seen the same way every time?

You've made reference to the album becoming the calling card for the piece. Could you explain that?

A recording if it's done well is an effective way of promoting a show. It's available commercially and has the imprimatur of being a commercial product. I don't think *Floyd* has sold more than maybe 30,000 copies but still a record company put it out and put out the booklet and all the songs and lyrics are there. I don't think the show would have had a life in the regionals and abroad without the record.

Myths and Hymns. **What was Tina's involvement?**

I had written a bunch of myths (many of which are not in the show) when I was about 23 and 24. The only really early one that's still in the show is "Icarus." It was a separate collection of songs, maybe six or eight. And then I found this hymnal in the old bookstore and those songs started to come out of me without a lot of warning. They were two discrete bodies of work. We were looking for something to do after *Floyd*; Tina came over and I played her a bunch of these things. She suggested on the second or third hearing that maybe we should do them as a song cycle—that maybe they would inform each other in a way that was not expected. I thought it was a really interesting notion. And as I began to write more, I realized they represent two systems for understanding our experience. Both are religious or "inspirational."

You say in the liner notes that one is what we aspire to be and one is what we actually are.

Or one is our prayers and one is our behavior. They inform each other and they are compatible systems in that way. Once that notion had been planted I started to write more based on that. For example, setting "Come to Jesus" into a little playlet—a little scenario in real life. That's what Tina's initial insight led me to.

At that moment we had a group of songs and we thought: "Maybe it'll have dance." Because after Floyd, where we got some reputation and kudos, we now wanted to have a commercial smash. And we got a little whiff of "well, maybe we could build on these songs and make a lot of money and some of these songs are really catchy. If we could just make a story out of this and find a narrative and get people to dance. . . ." So we raced through that briar patch for two years and ended up where we started, with a song cycle, because that's all the piece really should have been. The songs were not written together, they were sewn together. They happened to reflect on each other but only through a prism. There is no narrative structure on which to hang them.

What is your reason for doing *Light in the Piazza* now? Do you see a contemporary resonance?

I don't think I work like that.

What was your point of entry?

I wanted to write a love story. I wanted to find a vessel for romantic music. What struck me about the story is that it's a vessel for genuine romantic ambition and provided ways and angles with which to express those universal human traits. With *Floyd*, I think on a subconscious level I picked it to try and deal with the possibility that I may never be able to compose like my grandfather did, and I'm a fool to go into this business. With *Piazza*, it's that I have romantic ambitions like everybody else and I haven't necessarily attained them and was looking into that.

So it's really an exploration of love and romance from a number of different angles.

The way that I usually describe it is that the central relationship of the show between Clara and Fabrizio is like a benign virus that throws off a kind of infection which everyone gets. And this infection makes you come to terms with where you are on the romantic continuum. Are you at the bitter pole? Are you in the ecstatic first bloom of love? Are you somewhere in the middle? Have you been suppressing your dissatisfaction and sense of disappointment, and have you been burying it under a marriage? Everybody in this show is somewhere on that continuum and every song in the show references that continuum in some way.

How much material did you have when Craig Lucas came on board as bookwriter?

I'd written about half the score when Craig came on board—about six to eight songs—so I knew what I wanted it to feel like and what the tone should be. I use the word "tone" as a sort of a Swiss Army term—to me it means point of view. I had to decide what it was going to feel like and how it was going to move. How will it sound? How are we going to create empathy for these characters, an identification?

I also had a third of a script that I had written myself because for a minute I was going to try and do it myself. I think I actually wrote some okay scenes but I don't think I could ever do it all myself: it's just too much to keep track of, too lonely. I need another person.

What was it about collaborating with Craig Lucas that made the difference?

Laughter. It was fun to be in a room, it was fun to write. He made it fun to go to work. I had been feeling like a big failure. We went to the beginning and started again. We used a lot of my old material. Structure we figured out together—what a scene's going to do, what a song's going to do.

At what point did you decide that Craig would also direct the show?
We decided that later on. He really wanted to and I thought that we had discussed enough of the directorial concept.

How would you describe the nonprofit theatre as a place to work? In terms of your work, where do you see the nonprofit and commercial theatre in your future career?
Regional theatre is where I should develop my work—if they'll have me. Writers like me are going to the regionals and finding some measure of autonomy and decent money. A show has to have a distinct nature and that can be nurtured more safely out of town.

It also means that the show gets lots of different productions instead of touring one template.
They all get to remake it in their own image. I think that's great as long as the score is well sung and the story is well told. I think it extends the life of the show.

How do you feel about the increasing collaboration between the non-profits and commercial producers?
Nonprofits should remain nonprofit. They're entitled to a certain amount of subsidiary payback for bringing a show to life for the first time—they're taking a chance on an unproven piece. So the Intiman and the Goodman deserve a small share of the subsidiary rights of *The Light in the Piazza*. That, I think, should be the extent of their relationship with whichever producers come after them. Nonprofits fund-raise for a living and I think it is entirely appropriate for resident artists to help with that, to help pay for the creative safety they enjoy. Risk taking is the central mission of nonprofit theatre, which is protected by the patronage system and eroded by the commercial system.

INTERVIEW WITH MATTHEW WARCHUS
(May 20, 2003, New York City)

At the time of this interview, Matthew Warchus's new Roundabout Theatre Company production of Follies *was in the middle of its run at the Belasco Theatre on Broadway. Prior to this, Warchus had directed the musical* Our House *in London but was best known as a play director with his credits including the hit play* Art *in both London and New York.*

As a director who came up through the British theatre establishment and then tackled an American musical on Broadway, do you see any difference in aesthetics?

Possibly, a difference is seen when it comes to revivals. I think perhaps American revivals of classic Broadway musicals (and plays) might tend to be more conventional than British revivals of the same shows. Perhaps there is a freedom which comes from being "outside" the tradition which leads to a more revisionist, less literal, more "poetic" approach. When it comes to new works, however, I think the level of invention is equivalent between the two cultures (or possibly a degree more exciting in New York). What Broadway does fantastically well is this thing of energy and "presentation"—selling a number—which in Europe people don't do so easily. If you've got a show that requires that then it can be tricky in Europe whereas it is second nature for Broadway performers.

Why do you think that is?

Maybe it's connected to the American assertiveness and confidence—you know, "I'm here and this is what I can do." Whereas there's something more shielded and circumlocutory and protective and subtextual about British culture and maybe other European cultures as well. And maybe that is somewhat manifested in performance energy.

I enjoy spectacle and showmanship a great deal but also I expect my work on plays probably means that I try to find a way of making the relationships between characters in a musical really count, and try to encourage a style that is an unusual hybrid of interior and exterior. The interior is about absolute truth and documentary reality and emotion—authenticity and intensity—and the exterior is about abstraction—poetic imagery and spectacle. That hybrid is quite an interesting area. It is more usual for them to be mutually exclusive: people have either worked in areas that are essentially external or essentially internal.

Were there particular role models for your work on *Our House* in London?

Ironically, I thought a lot about the New York production of *Rent*. What I kept asking the actors to do was to "sell" the material but also try and keep hold of the psychological and emotional authenticity. The real trump card of that show is the energy of the young cast—like *Rent*—and I kept using the analogy of American performance style even for this very, very British story. One of the interesting things about Michael Greif's production of *Rent* is that a lot of the singing is done in profile or even upstage angles—more than in any other show I'd seen until then—and only a relatively small amount is sung out. It's very much people talking to/shouting at each other in song. I think this was rather groundbreaking at the time.

Your Broadway production of *Follies* ran into a very mixed and vociferous response. How did you find the experience of working on the show?
It was a real pleasure to do the piece as it's really intended to be done. Because I think it was ahead of its time when it was written. . . . Given that we [the Roundabout Theatre Company] were doing a show with a cast that large, the logistics meant that it was already way too expensive and the subject matter means that it isn't conventionally commercial. So we were forced to have a more scaled-down aesthetic—in other words, more inward-looking. Sondheim said that's basically the musical that he wrote, ideally to be done like that, in a book-driven way, with a strong, psychological, introspective, personality to balance the presentational "follies" set-pieces.

I believe that there were some rewrites for this new production?
To be honest, I think a certain amount of the dialogue was stylistically dated and needed to be modified. Some of the more obvious things could be said in a more understated way. I do remember doing a lot on what I call "Cacophony," which is when all four main characters and their younger selves are arguing. We wrote some new material in that bit and mixed it in with some bits from London [the 1987 production directed by Mike Ockrent] and some bits from the original. All the changes were about the dramatic shape of the piece and I hope we brought it to a more intense pitch.

The authors had given me all the original drafts dating back through eight years before the original show was written: there's some amazing stuff in there. In the original draft (it actually takes place in the Belasco Theatre) they go upstairs to the office and there's a sword fight between Ben and Buddy using swords that are taken off the wall in this antique study. And Sally comes in and shoots Ben. It's like *Uncle Vanya* or something because I think she shoots and misses: she tries to shoot Buddy. It's like black farce—the whole piece is so Chekhovian. I read the drafts and took stuff out of them. There was something about a baby—in that version Sally was pregnant by Ben when Ben left her for Phyllis. It was all very interesting, high-intensity stuff. Basically that cacophony was constructed by us in rehearsals using lines from different versions and also lines that we wrote ourselves. Other than that we restructured all the scenes of the reunion. In the dialogue when we meet all the different characters leading up to "Waiting for the Girls Upstairs" there are vignettes from different drafts.

Your casting of the show raised a few eyebrows, particularly in that the leads didn't have the big singing voices one would expect in a Broadway musical.
Of course, with Sondheim shows one is often looking for singing actors who can really mine the complexities in the piece. I thought the singing was good though. Have you heard the original cast?

When we cast someone like the wonderful Polly Bergen—who hadn't sung for 35 years—it was for all kinds of emotional reasons. There are people who could have sung it stronger vocally but *Follies* is a show that requires "dramatic" musicality and that is not the same as a big voice. Dramatic musicality is absolutely wrapped around emotion and the intensity of the show. You can go to shows and somebody can pin you to the back wall with their voice and that can be quite an experience but sometimes that is not enough.

The problem about a show like *Follies* is that it's attempting to do a vast thing: to look life absolutely in the face and describe an enormous truth, which is that Life is in many ways a continual slide from optimism and innocence to decay, and that everybody's life is like that and always will be in one way or another. In that sense it's about life being a certain process of dying. But it's also about being on that journey and how that journey is such a fast-moving one, and it's very easy to make a lot of mistakes before you've had a chance to think about it. And the road you're taking is informed by the number of mistakes you make, not realizing things in time—by not having knowledge, or an overview, about life until it's too late. So you get wiser after the event: you're losing your life but you're gaining wisdom. That's a vast thing to write a show about but that's what they've done. Like all great tragedies it's enormous. I just think it's an absolutely phenomenal achievement that they've made it as lucid and as entertaining as they did. Because the analogy of the end of the Follies is a great analogy, like *Merrily We Roll Along*, which has the same slide between losing your life/gaining your wisdom. I love the idea of doing it at a reunion where you look back over your life: the distance between the present and the past has been temporally a long distance, but emotionally it's just there, still present. Sometimes I think that *Gypsy* is the best piece of musical theatre structurally. But I think *Follies* is the best piece of emotional literature in the musical form that anyone's really achieved.

You called *Follies* a "tragedy." Could you explain that?
The sentence "I'm still here" is a sad line to me because she's not in life—she's still in this mirror of life, she's still lost in her anecdotes and her flirtations and wanting to be loved, needing to be loved. And that love—that addiction

people have for theatre, that fiction—is itself a symptom of a kind of dysfunction in life. Theatre is a refuge for people with a certain fragility or dysfunction. The fantastic thing is that that kind of dysfunction is alchemically turned into art through theatre. So the fact that some people can't be themselves—have to be other people, want to feel good about themselves—produces performances, which can then teach us about life. And that's healthy. Theatre's a great recycling plant turning dysfunction into enlightenment in a way.

Perhaps the problem is that the more right you get this show the fewer people want to pay money to see it. You know musical theatre attracts a lot of people who yearn to see an outsider or an oppressed character surviving, triumphing against the odds—it's a very cathartic experience. It's interesting that Sondheim is such an icon of musical theatre and yet his musicals are so at odds with that simple uplifting message.

You were originally slated to direct *Fiddler on the Roof* on Broadway. Why did you pull out?
Due to the intransigence of the Jerome Robbins estate and their concern that it was going to be different—even on the design front. Over and over again you come up against this dispiriting phrase "If it ain't broke don't fix it."

On *Follies* I was protected at Roundabout but conscious of just how many people had a sense of ownership in New York. It's very territorial and that was daunting. There's no real sense that these things are up for grabs and ready for refreshing. There's a sense that this is hallowed ground. And also the thing that Broadway does much more than the West End is that it makes icons of original productions. So a mythology springs up around things. . . . It was such a relief when the people I've gradually started to meet have said: "You know the original production of *Follies* wasn't perfect." Or even Sondheim saying it wasn't perfectly sung. Because all you're ever really told is that it was one of the greatest shows ever on Broadway. Often the mythologizing (and this is true of *Fiddler* as much as any others) is unavoidably entwined with commerciality. Because anything that's produced needs to be cloned and needs to be able to go on tour and needs to be around for a long time, so it needs to be mythologized and become an absolute landmark immediately: "There has never been anything like it!" "Won more Tonys than any other show."

It's interesting that a lot of Broadway shows now seem to be directed by British directors, even the musicals.
Somehow, British directors who have got a profile from other productions (and helped by Americans' misconceived idea of British being prestigious)

get a shortcut onto Broadway that perhaps some of the best American directors aren't getting. We achieved our profile in another country. To achieve a high profile within America I think you have to do quite conventional work, which means that the work you produce in the biggest arenas like Broadway is going to be more conventional and slightly old-fashioned. Of course, Julie Taymor is a complete contradiction of that. But rather like the British directors, she achieved profile Off-Broadway—in her own little country if you like—and then was allowed in like an imposter to do the same thing on Broadway.

It strikes me that all the British directors now on Broadway made their reputations not only in England but in other areas of the performing arts.
It's true that there is a mobility between different media, forms, and genre in the UK. It really can be possible for the same director to be given opportunities in opera, classical plays, new writing, musicals, films, and TV if it's what they want to do. Perhaps it's easier to take risks on people when the financial stakes are less.

INTERVIEW WITH WILLIAM FINN
(May 29, 2003, New York City)

*William Finn is a composer-lyricist who came to prominence through his "Marvin" musicals—*In Trousers, March of the Falsettos *and* Falsettoland—*which all premiered at Playwrights Horizons in New York. The latter two were later presented together on Broadway under the title* Falsettos. *His other work includes* A New Brain *and a song cycle,* Elegies, *which had just premiered at Lincoln Center Theatre at the time of this interview.*

Who were your influences as a writer?
Every day, before I would start writing, I would read 10 or 15 of Frank O'Hara's hilarious, moving poems. His diction—insouciant, conversational, and totally personal—reminded me of the kind of writer I wanted to become. Occasionally I would read Stephen Sondheim's lyrics before writing, but they were so good that they made me feel feeble, whereas Frank O'Hara's poems I always found welcoming and inclusive.

I'm always struck by the informal yet witty tone of your lyrics.
Witty? Yes. The informality, however, is a pose. That's just the voice I adopted; I have to work very hard to make my lyrics sound so informal.

How did you get involved with Playwrights Horizons?
Nobody would produce me, so I had to produce myself. I borrowed chairs from the temple across the street. Alison Fraser and Mary Testa would slice fruit and serve it to people and we were eventually attracting large, large crowds, especially for my workshop version of *In Trousers*, the first Marvin musical. They were real events. Andre Bishop had come to an earlier evening on the Upper West Side and then sent Ira Weitzman, the head of the new musical lab to see our little show. And Ira came up to me at the end of *In Trousers* and said, "We're starting a musical theatre lab at Playwrights Horizons and we'd love you to be the first person we showcase." And I said, snottily, "I've gotten a lot of offers and I have to consider them all." (I would have told me to go fuck myself, but luckily they didn't.)

I only went to Playwrights Horizons because the other theatres fell through. Can you believe this? And it happened that Playwrights Horizons was *the* theatre of the 1980s. And the people at Playwrights—Andre Bishop, Bob Moss, Ira Weitzman, Wendy Wasserstein, James Lapine, Rachel Chanoff—became my family. It was total serendipity.

Ira Weitzman seems to have played a vital role in your career and others in the 1980s.
Seminal. Nobody understands. He had the most well-developed ear of anyone I know. Plus, he was a meticulous line producer. He's the unsung hero of the new musical movement.

What was James Lapine's role in the *Falsettos* musicals?
On *March of the Falsettos* Lapine was instrumental. He brought a structural rigor to the piece that it had previously lacked and made sense out of chaos.

Did you find it easy to make changes?
At that time in my life, while I was writing *March of the Falsettos*, nothing was easy. Getting out of bed was traumatic. I was wracked by fear. So changing things quickly was not an easy thing to do.

Did you have trouble casting Marvin in *March of the Falsettos*?
When we did the original reading, I was Marvin, and thought any future Marvins should be just like me with a better voice. So Michael Rupert auditioned. And when Lapine asked if I wanted him, I said: "I don't know if he can do it, he's not anything like me." Lapine said, wisely: "That's not necessarily bad." So I went to see Michael in a small Broadway piece called *Shakespeare's Cabaret* and, having no idea if Michael would be any good, I Okayed him. I was not, however, prepared for his brilliance.

What exactly was James Lapine's contribution to the process?
He said "these are the songs you have" and he put them on index cards—
the title of the songs and what they were about. Then we got a board and he
started plotting. He said "This is what you want and this is what I want.
We're gonna go with what I want."

When he came into *March of the Falsettos* there was no child in the show,
no Jason. And he said "I work very well with children—can you put a kid
in it?" That's all he said. And I said, "Can it be Marvin's son?" and he said,
"I don't care." I came back to him with "My Father's a homo, my mother's
not thrilled at all" and he said, "now we've got a character." Without the
kid—who's the one acted upon—there's no show.

Then he said, "I think the psychiatrist should get together with the
wife." I said, "Do you know how offensive that is to psychiatrists?" So he
goes, "Oh. So they won't come to the show!" And in fact my psychiatrist
from Boston said, "Somehow I got over it by the end of the show but all I
could think was 'I cannot believe you had the psychiatrist marry his client,
what were you thinking?'" And all I could think about was "this show is not
about you, doctor." However . . . to get back to Lapine . . . On *March of the
Falsettos* he was very, very influential. *Falsettoland* was different because I
came up with the idea for the story. You know, there's a difference between
story and book. He would say "we need a song for Trina" and I would say
"well, I don't see why" and he'd say "well, we just do." And I said, "Well,
write out five lines about the song and let me see what it's going to be
about." He wrote out phrases and one of them might have been "holding to
the ground." I had no idea what the song was about and it really wasn't
helping me—they were just phrases. But it became what the play was about
without my knowing.

Did he plan that?
No, he was character driven—the character needed something. Or, rather,
the structure, which he intuitively understood, needed it.

Was there a lot of unused material?
I don't write that much material that I don't use. For *Elegies* I wrote, like,
two or three songs that we cut. But I had three songs that were ready to go
if we needed them so there were six songs that we didn't use, which is an
enormous amount for me.

Did you have a sense of the importance of these works?
Then—not at all. *March of the Falsettos* I thought was a simple bagatelle;
Falsettoland, the end of my career.

What did you not like?
I don't know. I don't know why I was so weirded out by it but I thought it was really bad. And I think it just has to do with my fears.

Not because it had been taken in a direction that you were uncomfortable with?
Not at all. I just thought it was really bad. And then we went for an interview right before the show opened. The guy says, "What's the show about?" and at the same time Lapine said, "AIDS" and I said, "Family." I thought "Oh, my God. We're working on two different shows." We looked at each other with horror . . .

There was some disagreement among critics about how far it was about AIDS.
The only way I could write it was to write it about family. I couldn't write about this horrible tragedy. What can you say about that? Consciously we never mentioned the word AIDS. And in fact some students saw it who really didn't know about AIDS but they were still moved by it—and that's why it was structured that way.

Did you see yourself as connected to a gay theatrical movement?
Not in musicals. I never did. Musicals it wasn't that way at all.

It seems to me that *Falsettos* is much more in line with . . .
I think *Falsettos* is much more important now than it was then. It was way ahead of its time.

. . . more in line with plays and playwrights at the time.
I had just seen [Martin Sherman's] *Bent* and the first act—I was so taken by that.

You've said in other interviews that you tend to write autobiographically.
I decided early on that I was a romantic writer and not a classicist and I was going to have to depend on myself for a lot of these stories. Now with *Elegies* it's very interesting where I've gone with it because it's become more and more personal. And I haven't yet found a place where the audience is not interested. Lapine told me the more specific I wrote the more universal it would be. And so I kind of take that as my credo.

What was it about *Bent* that had such an impact on you?
I thought it was so honest and revealing and gay in a way that I'd never. . . . Gay seemed so exotic to me, which is why I was able to write about it—it was like a foreign land. I was writing *The King and I*—it was

Siam to me. I could not believe I was part of this world, because I hadn't yet found my place in it. I didn't know what a normal gay life meant. So trying to put that in a play—I don't know whether that was hopeful or just trying to write a better version of what I saw my life becoming . . .

Were there particular playwrights that you admired?
I love John Guare. I liked it when you never knew whether it was funny or serious, those sorts of playwrights: Chekhov, John Guare, Wendy Wasserstein, Sondheim (of course), and Frank Loesser, who could musicalize anything. Musicalizing something inherently nonmusical seems a very dramatic action—arrogant, humorous, whimsical, yet serious. It says: "We are in the business of making the world sing." It's almost revolutionary.

Were your dramatic role models more from plays than musicals?
I don't know that they were even from plays after John Guare. . . . It was Frank O'Hara's poems; and Philip Roth was in the mix.

***March of the Falsettos* strikes me as a particularly innovative piece not only in terms of writing but also staging.**
I didn't think it was stageable. And I certainly didn't think it was stageable upstairs at Playwrights Horizons where there was no wing space. So because there was no wing space on one side and there was no airspace, Lapine thought to put everything on rollers and threw everything onstage like that. And the furniture was rolled around like chess pieces—which became a metaphor for the show.

How did you feel about the restaging of the song "March of the Falsettos" as part of the *Falsettos* double bill in 1992?
It was done brilliantly the first time. It never worked as well—Lapine has never been able to duplicate the staging. I don't know what happened and he doesn't know what happened either. It never looked as good as upstairs at Playwrights Horizons. Unfortunately, they didn't film that first version.

Did you have a different approach to the song the second time around?
We were trying to make too much sense of it actually. Originally, there was no fluorescence, and there was no coming up through the floor because there was no floor to come up through.

I notice that you tend to use the same actors in different productions.
My little repertory company. I don't have to explain things to them and I like writing for them. Also, I don't like that many people, but I like having them in the room. I more or less enjoy both their company and how they sing my songs. I hope I know how to write for them and they certainly

know how to perform for me. So, we're in the business of making each other feel talented.

INTERVIEW WITH DAVID LEVEAUX
(June 20, 2003, London)

David Leveaux is best known as a theatre and opera director having worked extensively in Britain, America, and in Japan where he is the Artistic Director of Theatre Project Tokyo. In 1996 he directed the musical Nine *at the Donmar Warehouse. This interview was conducted a few months after* Nine *opened on Broadway and the day after his new production of Tom Stoppard's* Jumpers *opened at the National Theatre in London. He was preparing for his new production of* Fiddler on the Roof, *which opened on Broadway the following year: the "discussion" that he was in with the Jerome Robbins estate was regarding the original choreography, which is contractually regarded as part of the show. Leveaux eventually lost his request for artistic license and the Robbins choreography was re-created within this otherwise reconceived production.*

I am interested in knowing what you make of the recent British-directed musical theatre "revivals" in London and on Broadway.
The key word that everything hangs on—whether negatively or positively—is the word "revival" because it's a word that you only hear in that context. I only became aware of it recently because when you come out of a tradition here—theatre—you don't think in terms of a revival as simply reassembling the shell of a piece.

Well, you don't talk about a revival of *Hamlet*.
You don't talk about a revival of *Hamlet*. You don't talk about a revival of *No Man's Land*. You talk about revisiting that play.

You talk about a production.
Exactly. So that there's always been an assumption in the British theatre that writing is by definition the map rather than the scenery—the map rather than the mountain. That the text or the book is a volatile thing and not a fixed thing in terms of how it can be expressed. But of course as soon as you get into that territory with musicals. . . . We're just in the middle of that discussion right now with *Fiddler on the Roof*. Jerome Robbins was contractually written into *Fiddler* as a writer as well as choreographer/director—but you know that if he was here, he would be completely ruthlessly redoing and reediting his own work. That's hard for people whose job is to protect his legacy to understand. But if you simply pick up something mechanically from 40 years ago

and dump it on the stage 40 years later in its original form, the audience has left you way behind. This is something that some agents often do not understand. And actually to be quite honest some critics do not understand.

Something I've been asked a million times is: "How does your *Nine* differ from Tommy's production?" When Matt Wolf interviewed me for his book about the Donmar, he said of my production there: "It's interesting because when you did *Nine* you cut Liliane's boa." Lilane Montevecchi famously did this number on the passerelle with a 30 foot feather boa. Fine—it was visual cue. And I said, "Matt, I want you just for a moment to imagine a 30 foot boa in the Donmar. Not only would it not fit but it would have no currency or relevance at all." But the idea that somehow one was cutting something from the piece itself simply by doing something differently from what Tommy did in the production of the piece—the idea that the original production somehow is the piece itself—is one reason why the piece hasn't been done for years and years.

I think Matt has one of those theories that goes something like this: the Brits do book-driven, text-based versions of the musical and the Americans do the more glamorous versions—they know how to do the spectacular musicals whereas the Brits pull it back from the spectacular into the more earnest. I know that deep down in that cliché there might be some element of truth but in reality it's not like that. *Dreamgirls*, one of the first musicals I saw on Broadway—that's the kind of musical I'd like to do. Michael Bennett actually took the musical on into new forms: that more naturalistic, semiliteral setting of musicals in the Rodgers and Hammerstein period was just transformed. He changed the aesthetic much more potently than any of the more recent shows. I thought *Les Mis* and that whole tradition that came out of *Nicholas Nickleby* with Trevor [Nunn] was a fallback to a kind of naturalism, whereas Bennett had gone beyond that and was coming out of a very different tradition of dance. And that was genuinely exciting and new.

Bennett was about rhythm, which of course all the great musicals directors are—they're about how you get in and out of a number, what the overall rhythm is. The problem with the naturalistic take on a musical is that you have to be careful that it may not supply the inner rhythm of the music itself. And that is first and foremost what people are responding to.

Why do you think that this surge of interesting British revivals happened in London in the 1990s?
Sam [Mendes] has a strong personal attraction to the American musical and a deep love of American culture. That's very strong in him. So when he opened the Donmar, the first show was *Assassins*.

Having said that, I think that there was another line of aspiration around at that time which gave rise to this, which is that all the directors you're mentioning [Trevor Nunn, Nicholas Hytner, Sam Mendes and Leveaux himself] became interested in *shows*. The generation of directors that we were brought up by (or sometimes "creatively abused by in a productive and challenging way"!) belonged to the period at the Royal Court where the political and aesthetic tradition was purity, which was very important in terms of stripping things down to the muscularity of language and the muscularity of dialectic.

When you got to the late 1960s/early 70s, people like David Hare, Howard Brenton, Howard Barker, and David Edgar wrote in a more fantastically decadent way. They were State of the Nation Plays. I guess Nick Hytner at the Royal Exchange started about the same time as I did and I think there was a conscious effort at that time to deliberately put more spectacular flair into shows. There was suddenly a kind of *visceral* excitement at what you were going to see. This is not to say that we weren't rigorous— there was absolute rigor and attention to text and seriousness about the responsibility to writers. But there was something theatrical going on after a period that you could call anti-theatrical (although many people would be furious at that term). The generation who came up with me had probably been exposed more to European theatre as well. I once sat in the Green Room at the National Theatre with John Osborne banging on about how disappointing the new generation was and I got so livid with him I said, "You know the trouble with your generation, John, is that you're so parochial you couldn't get on a hovercraft and go to Calais." And he goes, "Oh, you're probably right." So there was this European exposure through the World Theatre season. Suddenly there were these iconoclastic moments where someone would turn up on a trampoline doing *Yerma* and then they'd go back to their country. We became aware that there were other ways, other aesthetics—that it didn't have to be a choice between being anti-aesthetic and puritan about text or aesthetic and, therefore, careless about text. That it was possible to do both.

Nick came up in opera as well—I directed opera before I directed a musical. There was an interest in showmanship. I went to America in my early twenties and saw *Dreamgirls*, which just blew me away. The vitality and energy of it, the way light worked in empty space, the way Bennett operated the space and the dynamics of the musical form were as exciting to me as watching Merce Cunningham or Pina Bausch. There's a nonliteral push. The trouble with the musical is that it's actually a form that can express very serious themes but mustn't do it in a solemn way. Great musicals by their nature make these impossible leaps from one place to another in a way that you don't really come across unless you go back to *The Winter's Tale* or something. The beauty of "exit pursued by bear" is the imaginative leap. We all love music—and popular music as well. There's no shame attached to that. This idea that there's populist art and then there's musical art is starting to fold—as it should do because, as Peter Brook points out, the point about the theatre is that the divide that counts is between being alive or not being alive, not between "high" or "low" art .

Broadway producers seem to be choosing a lot of British directors. What are your thoughts on that?
I'm not quite sure why it is happening. What I know is that what happens with directors in this country is that it truly is taught as a craft—meaning that usually you grow up in a hands-on situation in the English theatre. Very few people move from theoretic directing or an academic course directly into directing. I'm not saying it can't happen, but what tends to happen here is that directors are allowed to start earlier, they're allowed to make mistakes, and they, therefore, quite quickly develop technical skills. The reason I say that is that irrespective of taste, sensibility, or the kind of theatre one director would make against another one, I think there is this quite old-fashioned thing that's taught here, which is that if you're confronting a scene and it's going a certain way you do actually have the tools to take it in another way as opposed to generalizing about it. I think there's probably something about the way that directors are brought up here that leads to a feeling that this is a job. So what happens sometimes is that a producer will decide "I'll go with a British director because I know that I'll get a show—for good or ill, there will be a certain show."

From an artistic point of view in terms of working in the theatre world, I don't think there is a distinction between British and American directors. I think Sam Mendes would say this, certainly I would. We think of it as us working inside the American theatre just as American directors (and writers) work here—John Guare and Jack O'Brien have been at the National Theatre and Joe Mantello was here with *Take Me Out*. I find that exciting.

On several occasions you have taken London productions to New York.
It's only with *The Real Thing* that we went abroad with the entire London company. And there were reasons for that because of the nature of the play. In the case of [his Donmar Warehouse production of] *Electra* there was talk about taking that production to the Public Theatre or taking it directly to Broadway. And, apart from the difficulty of getting the whole company there because of U.S. Equity restrictions, I also thought, "What does it

mean to take a British company (who I loved) in a play by Sophocles to New York?" . . . The last thing I wanted to do was to sit there as some sort of cultural event in the sense of "culture" in inverted commas because it was a "British" production. If we were going to do *Electra* on Broadway, I thought, then let's get it there because it is of itself an event and not some sort of "Brit" thing.

Nine was done on Broadway through the nonprofit Roundabout Theatre. Does it make a difference to do a show through them and not through commercial producers?
It does in this regard: in that Todd Haimes doesn't come and sit on your back and say, "I want this kind of show, this is what we're doing." He is actually very hands-off in that regard. His instinct is to function in much the same way as an artistic director of a subsidized or regional theatre would. At the National I don't have [artistic director] Nick Hytner on my back every day saying, "This is the kind of production of *Jumpers* that we want" but he'll give advice or see something and say, "Ok, my opinion is this. . . ."

But there was a specific reason why I took *Nine* to the Roundabout and that was to do with how to capitalize it. Given the expense of mounting the show on Broadway I knew that before we threw the switch on doing *Nine* I had to find a Guido Contini. I, therefore, never said, "We're going to do *Nine* at this particular moment." I just said, "Listen, for New York we need a Guido Contini and I think Guido Contini is Antonio Banderas. I have no idea whether he will do this or not or where he is." So that's how that started. A one-year commitment is basically the minimum that Broadway producers can go bat for but there was no way that Antonio could make an initial commitment of more than six months. Therefore, the only way that the production would be possible would be to massively reduce the capitalization costs. If you've got a nonprofit company that can operate on a different contract it keeps those levels down.

I went to Todd right at the outset and said, "This is going to be the most expensive project that the Roundabout ever mounted because we can't mount it for less." I wouldn't want *Nine* to be a charming cut-down version of a musical for us all to admire the book. The whole point of *Nine* is that it's a fantasia—you have to be able to go to *Nine* and feel enchanted.

Tommy Tune has described his love of the music but lack of belief in the book.
Tommy is a fantastic icon of an old-fashioned Broadway craft which I admire enormously. I will never make theatre like Tommy Tune—I wouldn't know

how to. I asked him to come to the first night of *Nine* and he came. And he said to me, "This is the first time this has ever made any sense to me."

What attracted you to *Nine*?
First of all it was the music. I just thought, "This is the most fantastic score." If you don't have a response to the music in a musical then don't even go there. And then the second thing was that it seemed to be driven by women and I thought that was very unusual and that was very exciting to me.

In a musical?
Well, to be driven by women of that nature, yes, I thought that was unusual. In this case 16 women. And I was excited about the notion of casting 16 principal women (as opposed to having an ensemble and then main characters) because of Fellini. I wanted those women to be not generic but specific and individual and alive like they are in Fellini's film or like they are in *La Dolce Vita*. Because if Fellini had done this musical, it wouldn't be an anonymous soubrette at the Folies Bergère—it would be Guido's mother. I was very excited to put women of genuine wit on the stage. Many wonderful actresses are too often given a choice—they're either playing sex or they're playing wit but they're rarely allowed to play both. And I thought, well actually this is about both.

I remember seeing Tommy's production and it was very stylish and cool. But now, looking back on it, I realize it was the height of the Gloria Steinem period and, therefore, the ways in which you could express women and sex were totally different. Whereas if we think back to the 1960s we discover that we can actually show women who are intelligent, who are witty, and who are also sexually mischievous—you don't have to turn them all into icons. We created characters—every girl had a story in the New York production. When people came in to read for it or sing for it I'd be thinking, "Oh, I see how she could be part of this world" and we'd invent a whole new story for each girl. It's not in the program note but there was one girl in *Nine* who, in her head, came from Naples to visit her boyfriend on one of Guido's movies and somehow the boyfriend didn't last but she stayed on.

Is there a specific reason why you set your production in the 1960s?
Yes, I think there is and I put it down to innocence. I don't mean naiveté, but innocence. The whole thing about the 1960s was that things were possible and the core of *Nine* is innocence. When casting Guido Contini you can't have a narcissist playing Guido Contini because the audience would just go like that [*snaps his fingers*]. This is a man who's a serial monogamist, he's not cynical. You have to believe that those women wouldn't be involved

with a man who was of that nature: they'd no sooner give him the time of day. And there is this kind of optimism and celebration of all kinds of different forms of love, some of which are admittedly probably not sustainable ultimately. But unless you've got that kind of glint of a world where sex and love are good things, as in the 1960s, then this is hard to express. . . . Post AIDS I suppose it's also hard to treat sex in a wholly abandoned and celebratory way, but I don't find it hard to believe in *Nine* if we go back to that world of the 1960s.

Also there was the staggering individuality: you could be anything, there was very little conformity. If you look at the way people dressed it's absolutely fantastic how daring that was. And so it opened up possibilities of individuality, iconoclasm—the whole sort of shammot of the 1960s. I can't see how *Nine* works in a modern context without allowing a certain amount of cynicism or exhaustion to enter in.

Of course, the updating worked very well for Sam Mendes in *Cabaret*. To create that world with the references to the 1930s unlocks that door and you've got something genuinely holding hands with the past. But the spirit of *Nine* is not that. It's in fact the other end of the spectrum. The way you hold hands with the past in *Nine* is having those women as ghosts. When we finally created that space at the O'Neill theatre, the thing I said to the designer Scott Pask was, "Well, we really should have a very ancient world with something very modern cutting through it as if the missing parts of the old world are completed by a completely modern structure—which is quite a 1960s thing to do. But it's ultimately so we feel that they're ghosts—these girls coming down that spiral staircase are like ghosts and instantly you see them it should open up a ball of nostalgia and loss in some way. And then the screens open up and we're going even further back to Botticelli: centuries of women, and that dance. . . . I just thought there's a dialectic going on here which is about longings, which is about something lost that's not entirely recoverable."

Nine is not a facile story about how at the end of the day you have to grow up and that those fleeting romantic relationships are superficial in comparison with the real one. The reason Guido Contini is a great director is because he loves women in a certain way and is able to film them in a way that reveals something—often quite radically. There is a genuine vitality and continually elusive light in those relationships he has with women, which are sometimes 30 seconds long. But are you going to say it wasn't worth it? No! Are you going to say that he learns that his life has been meaningless up until this moment when Luisa his wife walks back in to his life?

No! What you're going to say is, "Yes, there are things that we leave behind but that the reason that they're hard to leave behind is because they are in themselves glorious." I wanted people to bounce down the street after *Nine* thinking, "You know what, love is worth it, there's no need to be urbane or cynical about it, it's worth it even if it only lasts a moment. Women are certainly worth it, it's pretty good to be a woman and it's pretty good to be a guy around women." It's a big "Yes."

Friends of mine have been to see it and they get suddenly ambushed and moved by one moment in particular. This includes one woman, a film producer who's as tough as nails. The first time she saw it and saw the women coming back down to the little boy she said, "I just went and I don't even know why." And I think it's because of that thing—suddenly your guard is down and you say "I'm human" and you don't want that ever to go. So it's also for all those reasons that we went back to the 1960s.

Your production doesn't seem to shy away from the darker and more painful side of the show—or from the human cost of Guido's "freedom."
I don't think you can make a celebration in an authentic sense without that. For example, Carla's entrance for "Call From the Vatican" [staged as a show-stopping descent from the flies wrapped in a sheet]. The real reason I wanted to do that is that if you look at Carla, her sexuality is not actually predatory. If you look at what [the original Carla] Anita Morris did, that's the kind of sex goddess. The truth about Carla is she's on her own in a hotel room and the underlying gesture of even the "Call from the Vatican" is one of intrinsic melancholy, which itself has a kind of eroticism. In other words there's an eroticism that comes from that sort of sigh—a sigh that's actually very close to a form of sorrow. It's not that kind of "Oh, I'm being sexy" because that to me is not erotic. What I suggested to Jane [Krakowski, the actress playing Carla] was that if we could find that in "Call from the Vatican" then the journey to "Simple" becomes possible. Nobody vividly remembers Anita singing "Simple." They recall the stunning calisthenics in the body stocking [in "Call from the Vatican"]. Despite the fact that Jane would also stop the show every night with "Call from the Vatican" I think it was her "Simple" that really knocked your socks off—because it made sense in terms of the more "desolated" kind of Carla she was from the beginning.

There was always this possibility of sorrow inside the piece that's not in any way inconsistent with the notion of celebration. It's what I think makes a celebration worthwhile. What it mustn't become is sentimental or mawkish.

So her flying upside down in that sheet on her way out is also a pretty desolate image of a girl on her own. I made a joke on the first day of *Nine* that wasn't really a joke: "The thing is, what's got to happen is all the women have got to come from heaven and then go back there." I don't just mean that as a canonization of all women, I meant that there's something that Fellini does in the film—that extraordinary scene where this girl is pleading not to be sent upstairs and the kind of imminence of loss and of death. Fellini had to stick a note on the back of his camera—he taped a note in his handwriting to the back of his camera when shooting *8 ½* and it said, "Remember: this is a funny film." So that up-and-down thing is a gesture that operates on a wavelength of sorrow even when you're not actually thinking that. If we'd had the women coming in from the sides it would have been a completely different thing.

That's one of the fascinating things about how the theatre works—how the stage works—because sometimes you are operating at this purely intuitive level where someone doesn't understand why they cried. That's also why I love the musical as a form. It's not to do with getting solemn about text; it's to do with responding to what I would call the deep gesture of music, which is that music operates at all sorts of levels which are inherently dramatic if it's dramatic music.

For [*New York Times* critic] Brantley to describe Maury Yeston's score as being "mood" music is just so improper—no other word for it. The cunning of the structure of that music and the way it operates is that there are beats you don't even realize are affecting you because of the way that memory works. So you get to "Grand Canal" and what's happening is that all kinds of things are unraveling within that number—I mean, "Call From the Vatican" is in "Grand Canal." It isn't just about people suddenly going into a number. "Grand Canal" is a combination of the absurd, the clumsy, the aspirant, and the gorgeous—it's all those things in one brief masterpiece of a scene where this man puts a camera on his own life and you see women weeping and puzzled and enraged with him—as if he's absolutely lacerating himself for his treatment of them. And it's as if he somehow accidentally built this scene up from his own life. That's why I wanted to build this up bit by bit from fragments of what we had seen throughout the evening to this point, and to really emphasize with those images all the melodic refrains from earlier songs, so that it would primarily work musically. So you're looking at Carla with her bits of soggy paper sitting devastated in her chair, Luisa speechless, and everyone in his life absolutely horrified but somehow having to get to the climax of this absurd scene in a film about Casanova. And there on the back wall are the Three Graces and this music just going hand over fist building to an absolute disaster that has been waiting to come all night. And through all of Guido's life. That's why we did it that way—to express the catastrophe musically as well as emotionally. I don't see what's hard for Mr. Brantley to understand about that. But here's the thing. When Brantley says that bit's hard to understand he's actually trying to do two things at once and he can't. If I had reduced that scene to it's actual, if you like, plodding textual analytical meaning to make that comprehensive in the way that he's suggesting, he would be on me like a flash for failing to deliver the showmanship and the glamour. And he'd have been right. Because it's through that "showmanship" that we understand the story in this case. I set out to make a Broadway musical with *Nine*; I didn't set out to lecture anyone about the real meaning of *Nine*. I want to liberate what seems to me the very deep emotional gestures of that piece musically as a show. You can't do *Nine* if you don't like shows. Anyway, I love singing and dancing!

The opening image of a dining table in the production is a very communal image.
I was thinking: this is an attempt for him to gather together all the strands of his life (and essentially his life is measured by his relationships with women) and bring them all together at the start and then see how in fact he can't hold it together. What's happening in the first few moments of *Nine* is that he's looking for something. He doesn't yet know that he's fallen off a very, very high building. He's looking for something and in so doing he's checking back through his life. Each one of these people may have the answer to the thing that he's somehow missing. So I thought all those women would come to a place in his mind in which all those differences could be held together in a single gesture, which is a dinner table. Plus it's communal, plus the idea of coming together at a dinner table is fundamentally celebratory and is to do with relationships.

It's the equivalent really of assembling a small nuclear device, which of course when it reaches a certain critical mass does that [*mimes explosion*]. There is a stage direction from the original production about all the women talking at once and then suddenly Guido points for silence with his baton and conducts the women in the overture. I didn't buy that. I don't see how you can start *Nine* with an image of control. I don't think it has any relationship to the edge of chaos that is part of his life: the beauty of *Nine* is that here's a man who tries to make form out of chaos. I was excited at the idea of the overture being women continuing to speak but no longer hearing it as words. They're all saying something he's trying to hear.

You cut a number from the show: "The Germans at the Spa." Why was that?

I don't see how it fits musically into the piece. And furthermore, there is a dramaturgical problem. I said to Maury, "There's something wrong here structurally" and he said, "Ah, that's because I wrote it as a second opener because Tommy felt he needed one." Maury didn't have the slightest qualms about it being cut from the musical. The other thing is that if you've got a production like ours where you don't have generic ensemble—they all play distinctive individuals and never "double"—who are the Germans in that number?

The big dramaturgical problem is that you need to get on with Luisa's story and if you have "The Germans at The Spa"—a seven-minute number—by the time you get to Luisa you've slightly marginalized her. And it was very important to me that the real ticking clock—the real momentum of *Nine*—is not of a creative artist but a marriage. And that's why I didn't want to delay Luisa's side of the story too long at the start by having to wait beyond a number like "Germans." *Nine* begins with Luisa Contini saying to her husband Guido, "Are you listening?" and it ends in our production with Guido being *able* to listen. (Not with him racing after her into the distance. Because she has just said, "Be On Your Own" and she's not going to turn around. The only way you can preserve her intelligence is if you give her the choice—to *choose* to come back. Because *he's* finally got it.) Those are the bookends of the piece—it's a marriage. Without understanding that there's a marriage as the core of the piece then the other relationships don't really work—not least because there's also a sense in which all these women communicate with each other about this.

Claudia has that interesting line in the scene on the beach in which she mentions that it was Luisa who called her.

What's brilliant about that line is that it comes two-thirds of the way into the scene and we think, "Ah. Why has she held off telling him that until now?" And of course she's held on because there's a part of Claudia that's been thinking, "Will he finally . . . before I go back and say yes . . ." (to what I'm sure is the proposal of marriage that's been made to her by a man in Paris). ". . . will he finally say that he loves me?" And then she learns in that scene that it can never be and that, even though Guido doesn't understand yet, he's actually in love with his wife. And there's a release for Claudia in that. And so that's the moment she tells him about the call from Luisa. To try to tell him not to lose Luisa.

Antonio [Banderas, who played Guido] said: "David, I don't know—maybe this is crazy but I have a feeling that these women have a conspiracy between them. There's something they all know but he doesn't." And I said, "That's absolutely it." Because there is a strong sense in which he absolutely needs those women. Those women are in many ways fine with each other. They don't need this man; they choose to want this man. I think it's a brilliant insight that makes this musical very grown-up. It also liberates the women—it makes Luisa able to play herself from a place of strength not as a passive victim.

It also helps to emphasize the consequences: that you cannot keep living as he does.
No, you can't. It's interesting to me how women think and talk about these things, which is different from the way men do. It isn't actually as simple as competing. I'm not saying there aren't situations where that can happen—of course there are. But I was really intrigued by that area of them being able to acknowledge that there's something that they all share that brings them together—which is a choice they have all made in their different ways because there's not a single one of them who loves that man in the same way.

The casting seems to me to support that idea. There's something about having people like Chita Rivera in those ensemble roles that gives the women authority and strength.
Absolutely.

There was a new designer, Scott Pask, on Broadway. In terms of the set, did you rethink the Donmar production at all?
With Scott Pask, we gave him the idea of a table set in water, but he never saw the original. We worked together on the design for a long time. I think it was a great advance. For example, having the Botticelli on the back wall was something new. It really clarified that notion of being in an old space with a contemporary intervention—with the glass and the walkways. At the Donmar, the walkway came right round. And this time I went for something asymmetrical. At the Donmar we had two spiral staircases. There were things that we refined down and did away with. So ultimately that very tall spiral staircase had to stand alone.

It was a remarkable moment at the end when that slowly started to turn with the women on it.
We realized that we actually mustn't give that away and saved it for the last two minutes of the show. It's just for five seconds, but I love that five seconds because you can just feel an audience go "Ah!"

Applying this discussion more broadly, it seems to me that we need to get away from the idea of the musical as being one thing.

Yes, there are many different kinds and it seems to me to be an endlessly fascinating subject. Looking at it the other way round, *Jumpers* has a band onstage. There is language applied to that play which absolutely comes out of musical theatre. I think *Jumpers* is a play superbly disguised as an entertainment. I remember as a kid being taken to see the famous Peter Brook *Midsummer Night's Dream* and it blowing my mind when that started with two drummers with drum kits on top of this white box. And that's musical theatre too. The language we use in musicals is not dissimilar to the language we use in plays. I know some people who are in musicals like to tell you that there's a special science to musicals but actually I don't agree with that. The truth is that directing *Betrayal* is absolutely a function of rhythm: inner rhythm. Directing a musical: absolutely a function of rhythm. Maury said something funny to me about my having sprung fully formed from the waves as a musical director when really I'm a play director. The reality is that getting into "My Husband Makes Movies"—making it have dramatic momentum, exiting it and moving on—is totally the same language as working on Shakespeare. It's exactly the same. The problem is when people do it as if it's not. That's when you get into musicals where everything stops dead for a song. It's like that term "transitions" that people use: "The transition from that number to that number." And I say, "What are you talking about?" The transition is actually in the song. Perpetual transformation—that's how you make theatre. And the minute you stop transforming momentum is lost.

It's a matter of craft, I suppose.

Someone once said to me: "The thing about trees is that you can fly them in. The problem is that if you try to fly them out you'll get a laugh." Literally a few weeks after that I was in New York at a preview for *Sunday in The Park with George*: at the end the trees flew out and the house fell apart. That's craft!

INTERVIEW WITH IRA WEITZMAN
(July 30, 2003, New York City)

As a dramaturg and producer, Ira Weitzman has been crucial to the development of the serious American nonprofit musical. Starting out at Playwrights Horizons under artistic director Andre Bishop, he later moved with Bishop to Lincoln Center Theatre where he continues to work with writers on new musicals.

How did you get started with musical theatre at Playwrights Horizons?
I have known that my interest was musical theatre since I was a child listening, enthralled, to my parents' original cast albums. I saw *Company* a number of times as a teenager and that tremendously influenced what I thought was exciting in theatre. But I didn't know how that would translate in terms of a career. It was a pretty abysmal musical theatre scene in the late 1970s. There was hardly anything new being done on Broadway and there seemed to be no avenue for emerging writers. There was barely an outlet for established writers. And so I guess I had the chutzpah of a 20-year-old and I went to Andre Bishop and said: "Shouldn't we be doing musicals?" He loved musicals as much as I did—and was raised on them the way I was—so there was no hesitation to start moving slowly in that direction.

Bill Finn had been in New York for about a year at that point and had done a little showcase in his apartment of a piece called *Jocks* which Andre saw the year before I came along. Playwrights Horizons was not doing musicals then, so Andre said, "Good work, nice to meet you, so long" and that was the end of that. A year later Bill Finn was once again presenting something in his apartment as a showcase and it turned out to be *In Trousers*. As one of my first scouting trips I went to hear it and it was my fantasy of what I thought musical theatre could be: grown-up, sophisticated, and musically exhilarating. There was a character named Marvin but he really wasn't well developed—it was just a lot of brilliant stuff that was all over the place and unformed, but very exciting nevertheless. And I said: "Please come and work on it with us at Playwrights Horizons."

That was a definitive moment for me, for Bill, and for the theatre. Playwrights Horizons had just started a membership subscription program: I think they thought that doing musicals was something they could offer subscribers as a perk, like the little gift in the crackerjack box. But Bill Finn wanted more. He wanted to be treated the same way a playwright would be treated—with respect, with rehearsal time, and with proper support—not as a little showcase for a songwriter. Although he seemed like a very demanding, over-the-top guy, he really just knew what he needed as an artist and was not hesitant to ask for it and even demand it. And so it raised the bar for us in terms of how we thought about supporting musical theatre artists.

We would rehearse *In Trousers* in the middle of the night and everybody got paid $35—I don't mean per week or per day but $35 period. And the first time it was presented, it was for a few performances at 11 o'clock at night. Bill Finn and *In Trousers* was so infectious, his work started to permeate the halls of the theatre. Before long the whole theatre staff was singing his songs. That was incredible because it really started to show us the possibilities of doing substantial work in musical theatre, not just premiums for subscribers or showcases and cabaret.

So it became part of the theatre's main mission?
Yes, exactly right. The work was infectious right away but the commitment to musical theatre was more gradual. A year or two after *In Trousers* was done the charter of Playwrights Horizons was changed to read "to support the work of American playwrights, *composers, lyricists and librettists*." So it became part of the chartered mission and I knew that I had sort of arrived. Now in retrospect I think it was much more significant than just a little off-Broadway theatre making that commitment. I think it was the beginning of a movement, albeit gradual, of the nonprofit theatres being able to embrace new musical work the way they had embraced new drama.

It was a fantastic thing in my life and in the life of the theatre for Bill Finn to come along and, unbeknownst to him, lead us down this path. Bill Finn essentially writes autobiography. He doesn't write the story of his life per se but stuff that is inspired by personal experience in a way that musical theatre had never been allowed to do. Musical theatre on Broadway traditionally was a commercial venture looked on almost always as pure entertainment. Rarely were musical theatre writers inspired in the way that a playwright might be inspired to take from their own experience or their own lives and fashion that into a play. So Bill Finn and the first few shows of his that we did at Playwrights Horizons were significant in taking the musical theatre into that direction.

Lapine's work on *March of the Falsettos* and *Falsettoland* seems interesting as an example of the writer-director.
It's a cliché that necessity is the mother of invention. I don't think he was out to forge new territory or anything. Lapine had just started to emerge himself as a writer. He had a strong visual sensibility coming out of graphic design and he had a supporter in Andre Bishop. Lapine was around the community of Playwrights Horizons, as was Bill Finn, having been embraced by the theatre as an eccentric but lovable artistic genius. And it just seemed so natural that they should work together. Bill at that time was a one-man band. The first time we did *In Trousers* he played Marvin and he directed it—he wanted to do everything. He was an auteur. But it was clear that was too much for one person to handle. And so Lapine became interested in working with him.

Sometime after *In Trousers* we did a reading of *March of the Falsettos*, which was then called *Four Jews in a Room Bitching*. It was a wildly insane mess with no discernible story. It had episodes that were marvelous, it had songs and set-pieces that were intriguing but didn't really make sense. And so Lapine started to decipher the code that Bill Finn was writing in. From these inchoate vignettes he began to shape a family of characters and to make suggestions such as "What if they had a kid?" (Lapine always had a kid in his shows—it was one of his trademarks). The cast and everybody connected to it knew that Bill Finn was insane but brilliant and they had no idea what this could possibly add up to. Lapine would say, "Tomorrow I'm going to stage the part where the kid is upset that his father's a homo" and Bill would go home and write it. And the next day Billy would come in with the song. And that's how it unfolded. All the songs had index cards and it became a big puzzle on Bill Finn's wall. They would put the cards here and there and play with them and reorder them. There was no sense that they were creating a historic moment. It was all very fun at the time. But yet it made an enormous impact.

It got to the dress rehearsal and what had been an intriguing mess had become this coherent show. But not only that, it was magically staged. It started to define the chamber musical that the nonprofit theatre and Playwrights Horizons primarily began to champion because that was the size of our stage. That was the size of our budget. We couldn't afford more than five people so that's what we did a show with. That production was seminal. The chief drama critic at the time, Frank Rich, felt it was part of his job to make discoveries and to champion the new. And he loved *March of the Falsettos*, which put us on the map.

How do you see the role of the writer-director?
What Lapine brought to Billy was a writer's sensibility and a director's ability to carry it out. So in asking Bill to write material, Lapine would apply a writer's craft to it: "Ok, this is the moment we're leading up to; this is what I have to stage tomorrow. I don't have it in the writing—go home and write a song about this." So there was a writer's sensibility being filtered through the director's need to keep the thing moving.

So it was book writing in the sense of story and theme.
Yes, and in the end he received his due credit for that. It's an extension of what Hal Prince had been doing, though Hal is not a writer himself. What Jim Lapine did and what Tina Landau did with *Floyd Collins* was really applying writing skills to their work as a director. And so I think there really did start to be a change in the director-writer relationship.

Bookwriters have always been in short supply and as composer-lyricists like Bill Finn emerged, writing from a personal viewpoint, it made the task of a songwriter collaborating with a bookwriter all the more difficult. Because how do you find a bookwriter to write the story of Bill Finn's life— particularly if it's told primarily in song? So the kind of collaborator that Bill Finn needed to enable his work to come to fruition had to have a writer's ability and had to be able to work with him on that level.

What is the importance of the writer having a directorial vision from the start?

In the case of Bill, if Lapine had not been a writer-director but if he was just a stager he'd still be waiting for Bill Finn to deliver material! It was the writing sensibility that pricked Bill Finn to develop the characters, to make sense of the story, and to figure out what he was writing about. So if Lapine had not been a writer-director the production would probably have been less coherent than it ended up being.

Floyd Collins: **What was your involvement with that?**

I did not develop *Floyd Collins*. I have known Adam since he got out of Yale and was a huge supporter of his developing talent. We had worked peripherally and become friends and I often went to his loft to listen to the beginnings of *Floyd Collins*. I remember one of the first things he did was write this whole echo concerto—a big old 25-minute event. Adam and Tina Landau distilled it down to what eventually was seen on the stage. There was a great deal of experimentation on that piece. I was around for a lot of that unofficially because the piece was not affiliated with a particular theatre. By then I was known as the "friend of the emerging musical writers" so I recommended Adam to the American Musical Theatre Festival [in Philadelphia] to be a recipient of their Stephen Sondheim award, which came with an opportunity to do a workshop. The year he got it happened to be around the time he was ready to develop *Floyd Collins* and they chose to work on it in Philadelphia.

What do you make of the increasingly symbiotic relationship between commercial and nonprofit theatres?

There were very few opportunities for new musicals in the nonprofit theatre before we really committed to it at Playwrights. Broadway and the occasional musical at the Public Theater was it. In those days we eschewed commercial "enhancement" as we thought the art of what we were doing would be influenced in negative ways.

Nowadays it's hard to know what the role of the nonprofit theatre is in relation to the commercial theatre. I mean there are some blatant examples of shows like *Avenue Q* developed in the nonprofit arena but with commercial producers attached and a real eye toward its commercial viability. Today it's a daunting challenge for a nonprofit theatre to develop and produce a new musical without commercial support.

I feel fortunate today to be on staff at Lincoln Center Theater where we have the luxury of developing work for work's sake. If it happens to be picked up, transferred, and succeeds commercially that's a great dream come true but not necessarily the initial impulse. I don't feel that I do work for its commercial value. I still don't think that's part of our agenda. But in these times of dwindling arts resources one can't help but try to figure out what the merger of the two means for us. One of the most successful shows LCT has ever done was *Contact* and I assure you that it was done from a purely artistic need. Susan Stroman wanted to be protected in our basement rehearsal room doing her thing; she was not thinking, "I'm gonna make a fortune and do the biggest thing this theatre's ever had." It just happens that she has that capability and the show was well received.

Sunday in the Park with George: How did that come into being?
It came about because Jim Lapine had a commission as a writer from Playwrights Horizons. Sondheim sought Lapine out after seeing *Twelve Dreams*, Lapine's play that had been produced at the Public Theater. They started to toss around ideas about projects they might want to work on together. They discovered a mutual fascination for the Seurat painting "Sunday Afternoon on the Island of La Grand Jatte" and latched onto that as a fertile ground for a new piece.

We'd been developing and producing musicals for a few years by then but we had not worked with someone who we idolized as a musical theatre God and who had been a real established bona fide member of the Broadway world. We had been in our own cocooned world until then. Sondheim had been in the cocooned world of the commercial theatre until he did *Merrily We Roll Along*. Gradually it filtered down that Lapine was working with Sondheim, and that maybe we were going to do a reading of a sketch of something. We were reticent to even say we were working with Sondheim because everyone was very skittish about how to support this new collaboration. You know, we were walking on egg shells. But gradually over the course of maybe a year, maybe a little less, it started to emerge. The casting process began, we did readings, and we started to think about it as something that was actually coming to fruition. And as that was happening they were writing and writing. It seemed to be a very happy time for Sondheim. He had come off what must have been a low in terms of his Broadway work, because not only did *Merrily* fail artistically and commercially but it failed so publicly. Working at Playwrights Horizons was a safe haven for both Sondheim and Lapine.

Just to come full circle in our discussion, I think the fact that they were writing about the act of creation was an example of Sondheim embracing the notion that Bill Finn had of writing from the heart and writing from one's own experience. Not that Stephen Sondheim had the experience of Georges Seurat but he certainly had the experience of an artist working in a field where he was maybe going a little bit upstream. So the piece started to take on the themes of what goes into the act of creation: the pain, the joy, the satisfaction, the dissatisfaction, the relishing in the process as opposed to the product. The process of doing *Sunday in the Park* merged with the theme that was being written about. So it was a very, very exciting time for all of us. And it was the most expensive thing that we had done in musical theatre up until then. We took a lot of heat: "Why are you doing a Broadway musical in a nonprofit theatre?"

Was that because Stephen Sondheim was involved?
Absolutely. But this was not a typical Broadway musical. Where else, one might ask, should this show be done? There was this idea about the musical theatre being an inherently commercial endeavor as opposed to beginning with an artistic impulse. It's still true today, it hasn't changed. There are a lot of people who could care less about artistic concerns or serious grown-up matters in musicals.

INTERVIEW WITH JAMES LAPINE
(September 3, 2003, New York City)

James Lapine started out as a graphic designer and moved sideways into playwriting and directing. His Twelve Dreams *(inspired by a case study of Carl Jung) was first produced at the Public Theater and his breakthrough play* Table Settings *(about an eccentric Jewish-American family) was presented at Playwrights Horizons. He has collaborated with William Finn on* March of the Falsettos, Falsettoland, *and* A New Brain, *and with Stephen Sondheim on* Sunday in the Park with George, Into the Woods, *and* Passion. *At the time of this interview his production of* Amour, *with a score by French songwriter Michel Legrand, had played on Broadway the previous year and he was about to premiere his new play*, Fran's Bed, *at the nonprofit Long Wharf Theatre in New Haven.*

March of the Falsettos **opened in the same year as** *Cats*. **How aware were you of what was going on with musicals?**
I wasn't so aware. Not interested.

What sparked your interest in theatre?
Well, it was happenstance that I got into the theatre as I wasn't a theatregoer per se. I was mostly interested in artists like Robert Wilson and Richard Foreman and Meredith Monk and all the downtown world. That's what I came out of more than Broadway musicals.

Did you have an awareness of musical theatre?
Oh sure, my parents took me as a kid. I was from a small town in Ohio, home of the lyricist from *Bye Bye Birdie*, Lee Adams. We went and saw that. But when I became an adult it was too expensive and I wasn't that interested. It seemed like golf to me: kind of boring.

What was boring about it to you?
You have to put it in the context of my era, which was very counterculture. In the 1960s and before that so much of the contemporary music was the theatre music as well. You had Frank Sinatra doing cuts from shows and everyone owned their copy of *Camelot*. But my generation's music was folk and rock and acid rock so the Broadway music suddenly seemed old to me. It represented what everyone in the counterculture was desperately trying to get away from.

You've done both plays and musicals; you've been a writer and director of both; you've worked in nonprofit and on Broadway. Is there a particular strand that you see running through all your work?
No. A lot of it has to do with people. I've worked with Andre Bishop in the world of nonprofit theatre. Then when I got involved with Sondheim, who's sort of a commercial guy, it led me in that direction. There wasn't any kind of determined effort to go here and there—it has to do with who you're working with and the kind of projects you want to do and what dictates their venue.

Would I be right in thinking that the psychological angle is often a point of entry for you?
I'm totally not self-analytical when it comes to "career" so you can think whatever you like! I don't analyze what I'm doing, I just tell the stories that interest me.

How would you describe your work with Bill Finn on *March of the Falsettos*?
I'm a structuralist and Bill is . . .

. . . not?

At the time I wouldn't have even known I was a book writer but essentially what I did was to come in and write the book of the show by expanding plot and characters. Help open up his story. I suggested the role of Jason, for instance.

Why was that?

I like working with kids and I just thought it would humanize it in a way. Bill's work is very edgy and in your face and I thought a kid would be a nice kind of counterpoint to what was going on.

But the thing about collaboration is that it's ineffable. What was nice is that I was influenced by the more avant-garde, process-oriented school and in those days you could do that: just say, "we're gonna put on a show" without having a show. I think it's much harder to do that now. The process was kind of unencumbered—there was no expectation. And when you're at that age, too, you're very fertile, you really have a lot of ideas that just race through you that tend to die out a little as you get older.

Mostly *March of the Falsettos* came out of a certain spirit of discovery and adventure. It was special because it wasn't plotted out. The set came together day by day. You know, every day it came in focus a little bit more like a photo. I think *March of the Falsettos* has the spirit of it. If you look at the sequel 10 years later it's a much tighter, much more mature piece of work. I worked on a movie adaptation and it's difficult because you don't want to lose the enthusiasm of *March* but on the other hand it's a little (Bill would be the first to agree) it's a little sloppy. The thing about *Falsettos* is it was just not like anything else out there—it was quirky and it had the benefit of an incredible score that was very accessible, and a topic that no one had written about.

It must have been challenging to work with someone who writes so autobiographically.

Well, Bill's writing is not so much autobiographical as personal—idiosyncratic.

You've said elsewhere that you had some problems with the original writing of Trina. Is that something that bothered you at the time?

It bothered women at the time. I was very tuned in to the fact that she was the only woman in the piece and it was also that time when everyone's antenna was attuned to how women were portrayed. The nice thing about Bill is that he doesn't think about things like that. He just writes what he writes, he's not P.C. and doesn't pander to certain expectations and that's one reason why his work is so unique. I think that character developed, partly due to different actresses playing it.

Did you see yourself as a cocreator on the show? Or as helping Bill Finn to work out his ideas?
You know, I didn't really think about it. I think if you collaborate on something you are creating it together in a way. It depends. Certain writers give you material that's completely mapped out and drawn out and your job is to bring it to life, whereas other writers give you a less complete piece to begin with.

What was it that you found interesting about this project?
I found it intimidating actually. I remember being anxious about doing it because it was edgy and in your face. You have to remember that back then a gay-themed musical was pretty audacious and *March of the Falsettos* dealt with a lot of hot issues. I wasn't worried about what it was going to turn out to be so much as the fact that the process was kind of scary.

Did you approach Bill and say you were interested?
No, it was just the opposite: Bill sought me out. He was persistent about my doing it. I don't even remember deciding to do it, it just sort of happened. His enthusiasm was contagious.

I imagine that he can be quite persuasive.
Yes. He's scary too in his own way. He's such a force of nature. An extremely smart and talented fellow.

Did you see your function on *Falsettoland* as being different?
Well, it was much easier to do because we knew who the characters were. It was an interesting thing to do too because we were writing a sequel that's set a year later and yet it was ten years later that we were writing it. So we were writing with hindsight. . . . It was much easier to do because we knew who we were writing for and we had that big dramatic story to tell.

Would I be right in thinking that you don't work thematically—that you approach shows from the angle of specific characters or situations?
Yes, I don't tend to go for big overviews. I tend to like it to evolve from what's at the centre. I'm not all that conscious of what I'm doing. I work more instinctually.

Ten years after *March of the Falsettos*, what was the impetus to go back and do another musical about the same people?
I think it was mostly about, "Gee, that was such an incredible experience, why don't we do another?" I think they're interesting characters and I think

those people and those stories should never stop, just keep going forward. Besides, we had a 10-year perspective on the AIDS epidemic and that was a subject worth exploring.

I believe it was Graciela Daniele who first put the two shows together as a double bill.
Well, we had wanted to do that initially but we couldn't get anyone to do it.

When you came to put the pieces together did you find that you had a different approach to the piece?
It wasn't hard to put it together because I had already done both of them. The only thing that was difficult was that there had been a great review of the Hartford production. But I felt lucky to do it in NYC and I thought it was very magnanimous of Grazie to graciously step aside. She had done it on a thrust stage in Hartford and the original plan was to move it to the thrust stage at Lincoln Centre. That fell through. And then when they were going to do it on Broadway Graciela didn't feel she had the time to reconceive it for a proscenium stage. There was a lot of tension at the time. All the people who were in the original, myself included, had very mixed feelings because we'd worked very hard for a long time for very little money and it was finally a chance to see some return. So it was an awkward situation and Bill was in the middle of it. But it all worked out in the end.

With revivals of shows that you have cowritten, do you have a strong feeling about where it stops being a new interpretation and becomes a distortion?
When you're fortunate enough to have a show like *Into The Woods* that's had so many productions it's like an opera—once they're established then you can take lots of liberties. But I think you always have to serve the material and the problem is when people get almost diabolical in sabotaging material and basically rewriting it to serve their own vision.

There seems to be an underlying issue of the original staging being treated as part of the show. Do you have any feeling about that? Is the show what's on the page?
Yes, I think it is. The unfortunate thing is that Jerry Robbins is such a genius that I've rarely seen people improve on what he's done. Directors will often do something differently than you did just so that they don't do what you did. But I suppose you have to give people that option. Yes, I think in time people's memories of Jerry Robbins' productions of X, Y, and Z will diminish and everyone will reinterpret these works.

But I must say it's so much easier to do a revival than an original musical. It's much easier to have seen something and seen what didn't work. When you're doing something for the first time you have no objectivity—you're fighting so hard just to make the material work that often you don't have the option of going back and saying, "Oh God, I wish I hadn't designed this or cast her." You're at a great advantage doing a revival. You have a framework and you can steal what you want to steal.

Moving onto the Sondheim collaboration.
I initially met Steve about doing *Merrily*. It was George Furth who had moved to have it revived. But it had just closed on Broadway so it was way too soon to do it.

You have previously said that Sondheim taught you a lot about writing musicals. What was it that you gained from him? And what did you bring to the partnership?
Well, Steve's very open so it's not like he was entrenched in a style of work-ing. I learned a lot about structure and songs' purpose. And I think the whole process is what was exciting for him—the process of putting shows up before they're written; having a workshop. I think it was exciting for him to do theatre on a small scale.

I got the impression that you were very instrumental in bringing in ideas.
It kind of evolved and I had used that image [Seurat's "A Sunday Afternoon on the Island of La Grand Jatte"] before in a little show I had done. Then I started writing and we would talk about the writing and eventually it just built itself into a show.

At the ideas stage did you also bring in other sources such as short stories or was it just images?
It was just knocking ideas around. I don't think it dawned on me to adapt something; the idea was to create something original.

When you were discussing what to write about was that in terms of "this would be a good topic to write about" rather than "here's a story idea."
It didn't come that way. It really came from being intrigued by an image. I don't think Steve had any idea what it was going to be until I brought in the first six pages. He didn't know where it was going and a lot of it we didn't even discuss. I would bring in six pages and then another six and then another six. Then we started to hammer out where to take it next.

How far did you go with your writing?
I wrote basically the whole first act without his writing anything, though he wrote the intro chords and thought about where he thought a song might go and what songs he might want to write. But mostly I wrote the first act.

Did you discuss the story and characters as you went along?
Some.

But normally you would write something and then you would talk about it afterwards?
Right. There was no outline. And then we talked about what any subsequent acts might be. It's hard for me to remember how things evolved.

Did you ever consider just leaving it at that first act?
No. It was always meant to be a meditation on the life of a painting. It had initially been three acts that tracked the whole life of the painting but it was so ambitious that we decided to take the second act out.

What was that?
It was what happened to the painting—which was basically that it got rolled up and stored in an attic. It was eventually bought by some Americans for a song and brought over here. Also, a few of his paintings that followed it were tracked too—there were other paintings that were enacted onstage as well so that we got to see all of Seurat's major work up until he died.

Did you actually write that?
Yes, I think we did write it. I don't remember at what stage we gave it up.

How much of what we saw on stage was Tony Straiges and how much was your design?
A lot of that was fairly literally in the script but obviously Tony's the one who brought it to life.

So the use of paintings onstage and the use of those visual effects—that was in your libretto?
Yes.

The London productions of *Sunday in the Park* weren't directed by you. Was that a conscious decision?
Yes. I don't remember whether they asked me to do the first but I wasn't interested. I don't find it interesting to go back to old things. You know, once you've done it to then immediately go over it again with new actors is really like retooling a car. It's not that interesting a process.

What did you make of the first London production?
I didn't have a good time. It was kind of a shock. I remember liking Maria Friedman quite a lot and Philip Quast. I hated the second act—I remember they did some weird business and they sort of rewrote the whole thing. It felt like it diminished the character. I quite enjoyed the latest revival.

Was *Into the Woods* a different kind of collaboration with Stephen Sondheim?
Well yes, because we knew one another.

Was this another case of you writing material and then going back to Sondheim to talk about where songs would fit?
Well, sometimes I would suggest a song. Sometimes he would say, "I want to do a song about X" and I would help write a scene that would go to a song about X.

But the initial step was presumably to sit down and work out which stories to use?
Well, *Woods* is much more plotted so I think we had a lot of conversations about plotting and outline. It was very different than *Sunday*.

And that was something that you both did together at the initial stages?
Yes, generally. I would usually do the groundwork.

In terms of drafting?
Yes, drafting it or suggesting plot scenarios. Sometimes I wouldn't talk about what I had in mind. I wanted him to read it first.

And at what stage did the songs come?
I remember writing the first scene thinking, "There's no way that this can be musicalized." And I remember Steve finding a way to musicalize it—I thought that was pretty incredible. It's a complicated show—it's maybe too complicated.

Would I be right in thinking that Sondheim is more of a structuralist than Bill Finn?
Oh totally. Steve writes very slowly and doesn't want to write a song until he knows exactly what he has to accomplish. Bill will write you a song very quickly and then refine it.

I heard that with Bill Finn you could ask him to go home and write a song overnight.
Yes, you can't do that with Steve. You don't want him spending three weeks writing the wrong song.

Did you see the Richard Jones London production?
Yes, with the doors. Bizarre. I enjoyed it. I mean, I thought it was not user friendly to kids—it was very dark. I have to say I really enjoyed it. He's quite inspired, Richard Jones, with a bizarre take on the world.

And it didn't bother you that they did it in such a different way?
Not at all. I loved it. I thought it was totally unique and kind of brilliant.

With *Passion*, you worked from an adaptation. Was that a different process?
It's funny because Steve was attracted to the movie and then I read the novel and actually really loved the novel. It's much easier to adapt something because you have a common source—you know the characters, you've seen it, you're working off the same notion. So from that perspective it was easier to do and maybe a little less interesting. The novel had a lot of great detail in it. The movie was much broader—that characterization was so extreme. I remember reading Vincent Canby's review of the movie and him talking about it as a comedy. And I thought, "Oh my God, I really didn't think I was watching a comedy."

So when you came to turn that into a musical did you try to get the novel's level of detail into it?
Yeah, and I think it was helpful to Steve too because there was so much more character detailing in the novel to draw on and fill out. The hard thing with that show was that when we put it up at first it didn't work. It worked in a workshop form but it didn't work to a general audience because it was difficult to get a commercial audience to buy that journey—why this handsome guy would fall in love with this unattractive woman. Steve and I got it. We had to work in previews to help others understand.

A reviewer in Washington DC recently said that it clicked for the first time in the Kennedy Center production there.
Well, I think it's a show that is helped by being seen more than once. What's so ironic is that with Sondheim's work in particular you get people going "this revival is better" and it's like: "Hello—you're seeing it for the third time! You've listened to the CD." That's the point of seeing something over and over or reading a book twice or going back to a painting—you get to see things and hear things you hadn't seen or heard before. If you hear or understand something on a second viewing it doesn't necessarily mean that it's because of the production.

It's funny, I know a lot of people liked the Washington production very much. I didn't care for that Giorgio personally because I felt he was so vulnerable and such a sitting duck from the get-go that there was no story. A lot of people liked that about it. So it's a matter of taste. The best thing is that the piece is strong enough that it can withstand that kind of interpretation. I'm thrilled that they liked it and that it went well. A few years later I directed a reading of it with Michael Cerveris who did the DC production. I asked him to take it in a different direction and I thought it was sensational. Of course, I have no objectivity!

I saw the 1996 London production directed by Jeremy Sams, which seemed to me more about Giorgio's journey.
Yes, well they added that song ["No One Has Ever Loved Me"] in London which they also did down in DC. Dramaturgically and musically—on every level I didn't like that song. I think the shift had a lot to do with Michael Ball being in it.

How did you feel about that production?
I tried to keep an open mind. I like Jeremy a lot. I wouldn't say that my production was in any way definitive so I didn't have a problem. In a way *Sunday* felt definitive to me in a way that nothing else has. So I didn't really have a strong feeling. I loved Maria. They put in an intermission. You know, let's face it: it was critically successful here and won a bunch of awards but it wasn't commercially successful. So I thought, "Somebody has a better shot at it—do it! What have I got to lose?" In the end it didn't do as well over there. The problem with *Passion* is it's a small show and it was never meant to be in a big theatre. Probably the best production was the production at the Signature Theatre [in 1996, directed by Eric D. Schaeffer], which was 150 seats with a 12-piece orchestra. What could be more perfect?

INTERVIEW WITH MARGO LION
(April 2, 2004, New York City)

Margo Lion started out in politics and then moved into theatre producing, first within the nonprofit sector and then as a commercial producer. Her Broadway credits include plays (Angels in America, Seven Guitars, *the 2002 revival of* The Crucible), *musical comedies* (Hairspray, The Wedding Singer), *and musical drama* (Jelly's Last Jam, Caroline, or Change). *This interview focused on the origins and development of* Jelly's Last Jam, *clarifying and expanding on the material already set out in Marty Bell's book* Broadway Stories.

Did you have a strong idea of what you wanted with *Jelly's Last Jam* when you started the project?

My goal was to produce a musical that told the story of how jazz came to be: how this extraordinary art form emerged at the turn of the last century in the city of New Orleans. I wanted to find a good story, to see this moment through the lens of a compelling central character. When I read Alan Lomax's interviews with Jelly Roll Morton, the self-proclaimed inventor of jazz, I knew that I had found my man. Here was a larger-than-life figure whose intriguing biography and tuneful melodies along with his revered place in musical history made him an irresistible subject.

You've said that you were to some extent inspired by *Ain't Misbehavin'*? What was it that you liked about that show?

Through the arrangements of the music and the order and staging of the songs, *Ain't Misbehavin'* evoked the subtext of the urban black experience . . . the joy, the loss, and the determination to endure made for a memorable evening in the theatre. It convinced me of the theatrical potential of a jazz score and left me wondering how this uniquely American sound was born.

I believe that at one point August Wilson was attached to write *Jelly's Last Jam*?

My decision to ask him to write the book was misguided. I was a novice producer who saw August as a unique chronicler of the African-American experience but as it turned out, he had seen only one musical in his life, *Zorba*; he had no familiarity at all with the form. The book for a musical is less about literature and more about structure and economy of style; there was no room for the gorgeous language and unhurried pace that characterizes August's writing. By mutual agreement we parted ways after a year of trying to tease out a workable first draft. Happily, we continued our working relationship when I had the good fortune to be a producer on Wilson's *Seven Guitars* several years later.

What was it that George C. Wolfe brought to this show that other African-American writers had trouble with?

Jelly Roll Morton was a New Orleans Creole and on occasion an outspoken and even vicious critic of black Americans. The first playwrights I approached to write the musical reacted viscerally to Morton's attacks on black musicians and rejected my proposal as perpetuating a negative stereotype of African Americans. It was only when I found George Wolfe that I discovered a writer intrigued by the dramatic potential inherent in the conflicts and prejudices that existed inside the black community. His vision was far more expansive and compelling than my original notion of keeping the musical in New Orleans and focusing on the less complicated and more surface entertainment value of Morton's music and his self-aggrandizing and colorful persona.

Whereas Wilson celebrates African-American culture . . .
Jelly's Last Jam is certainly a celebration of African-American culture. After all, it is only when Jelly acknowledges his black heritage and how fundamental that heritage is to his music and to his own identity that he is redeemed and gains admission to the pantheon of jazz greats.

George comes from a nonprofit background and there has clearly been a recent shift in the commercial-nonprofit relationship. Was this a project that needed that relationship?
Jelly could only have been developed in a partnership with a not-for-profit theatre. Given that the show was an exploration of a highly complicated and unconventional African-American central character, one who was not easily recognizable to a traditional Broadway musical audience, George needed the opportunity to work through the material in a protected environment and with an audience that was used to seeing new work. Of course, given the fact that this was George's first outing as a commercial director and that he had never directed or written a musical (his other hat), there was little hope of interesting commercial coproducers without an earlier production with good reviews and positive feedback.

As it turned out, we needed one more workshop and the addition of Gregory Hines to the team to make it all the way to The Virginia [now the August Wilson] Theater. Gregory was convinced by the power of the Mark Taper Forum production to come on board; without that first look, he may well have passed on coming to Broadway leaving the future of the show in jeopardy.

Do you still think that there are restrictions on musicals in nonprofit theatres?
Nonprofit musicals can be as daring as they want as long as the storytelling and the execution are compelling. It's obvious that there is far more latitude with subject matter in a subscription season than in an open-ended commercial run. The major concern for a subsidized theatre producing a musical is financial.

Is this a bigger factor with musicals than with plays?
As musicals are very expensive to mount, they inherently require greater access to funding than plays. When nonprofit theatres choose to present

new musical work, they will most probably need significant enhancement by a commercial producer.

INTERVIEW WITH JACK VIERTEL
(April 5, 2004, New York City)

_Jack Viertel was a theatre critic for the Los Angeles Herald Examiner and sub-
sequently worked as a dramaturg at the nonprofit Mark Taper Forum in L.A.
As the creative director of Jujamcyn Theaters, Viertel is now involved in finding
and producing both plays and musicals for Broadway. He conceived and copro-
duced the musical revue_ Smokey Joe's Café, _was a dramaturg on_ Hairspray
and cowrote the musical Time and Again. _Viertel is also the Artistic Director of
the_ Encores! _Series, which presents big-budget staged readings of "lost" musical
theatre classics at New York's City Center._

**What is your sense of Hal Prince's role within the development of the
musical?**
He was a pivotal figure. He was a complete master of the traditional musical as it was crafted by George Abbott and Jerome Robbins—he wasn't
someone who came from the serious world and tried to make a musical.
And yet he had more serious ideas that he wanted to explore, and in
Stephen Sondheim he found a partner to explore them with. He set up a
model for what the musical could be about—the kind of subjects that it
could treat—that was fresh. People had from time to time tried to do that
but I'm not sure anyone synthesized Broadway musical theatre technique
with more advanced ideas in the way that he did.

   He did it in the middle of the Vietnam War, which is interesting because
it very much grew out of the restless mood in the audience—things like
Company, Follies, and _Pacific Overtures_ were analogues to the mood of the
country. They weren't literally about Vietnam and disaffection but they
were very much from that national mood.

**There were of course earlier periods when musical comedy was a
suitable reflection of the national zeitgeist.**
Right. And for the most part it seems to me that the previous serious musicals like _Showboat_ and _Pal Joey_ and _Porgy and Bess_ were anomalous. They
existed in a world that was nothing like them. _Pal Joey_ didn't change anything. At the time, it was treated as a serious step for the musical but it
didn't really have any wake. It didn't portend a shift in the way that shows
would feel for the next few years. But when Prince came along that really

did happen to some extent. Of course, Rodgers and Hammerstein created a more serious musical theatre in the 1940s, and were joined by some others, like Kurt Weill. *Oklahoma!* is a fascinating treatment of the idea of statehood, and *Carousel, South Pacific*, and *The King and I* all qualify as serious musicals and probably laid the groundwork for the kind of ambitious work that Hal Prince did. But they didn't really question the world in the way that he did. Hammerstein, who is the major architect of the American musical, was basically a positivist, although not a Pollyanna. Prince came from another world, and he began to mistrust basic assumptions about people and morality in a new way.

Starting with *Cabaret* . . .
Starting with *Cabaret*. That was a big surprise. I remember *Cabaret* very vividly—I was maybe 16. Hal had directed *She Loves Me* before that and *A Family Affair*. *She Loves Me* was remarkably integrated and wonderfully done but it was a souffle, it was a wonderful romantic story. And then suddenly this dark thing came along.

Do you have any thoughts about the idea that *Follies* represented a tussle between Prince and Bennett?
I don't agree. I've seen *Follies* approximately 40 times and it didn't feel that way to me at all. I actually know a little about it: I was at college in Boston when it was trying out so I used to go and see it every three days and see what they were up to.

Even though Michael Bennett was credited as a codirector—and no doubt did a great deal of it—it felt like it was essentially Hal's vision. The "Mirror" number was a great tap number, but it mainly took your breath away because having the eternal idea of facing up to oneself in the middle of a tap number was an extraordinary, unique event—and it felt very much like a Hal Prince idea.

The show was almost completely joyless. I loved it but throughout the run people were getting up and storming out from about an hour ten minutes in. It was a completely fascinating deconstruction of the death of the American Dream if you watched it that way. Even things like the trio of songs (*Broadway Baby, Ah, Paris* and *Rain on the Roof*): each one of those deals with a different kind of American dream that's gone sour. But by and large audiences didn't read it that way.

It felt unsolved in the sense that there was nowhere to go in that show. This guy had fallen in love with one girl and married the other and it was just a tragedy. So any ending that they wrote to convey a ray of hope would have seemed unconvincing. There's a great chapter in William Goldman's *The Season* about The Three Theatres. The terrible mistake is to pretend you're doing one when you're doing the other. *Follies* did it intentionally. It is basically Third Theatre—hugely expensive Third Theatre and presented so it looks like popular theatre but it really is Third Theatre, an extravaganza designed to bring you up short instead of reassure you.

It seems to me that if an innovative show like *Follies* were to be done today it would have to be done by a nonprofit theatre for financial reasons.

Yes, but what *Follies* had that you could never get at a nonprofit was Hal Prince's innate sense of show business. It was deceiving because it felt like a great big Broadway musical but it had no joy to offer you—that was its signal achievement. Even the mirror number that made you want to scream with admiration didn't actually make you happy—it made you feel horrified. My experience with *Follies* is very personal because when I first saw it I was the same age as Young Ben and I'm now older than Old Ben and Buddy. When I went to the 2001 revival I completely broke down at the end of "Beautiful Girls" because those look like the women I know. When I first saw the show I was only 21, and when they came down those stairs in the revival my whole life passed before my eyes. It had that kind of power.

Moving on to the so-called British Invasion of the 1980s and 90s.

I was in LA at the time. I was a theatre critic.

Would I be right in saying that there was a sense of resentment about the British shows? A sense that whatever else was happening in the theatre, musicals belonged to America and to Broadway?

I think that on the street there was a combination of resentment and admiration. To be fair, Americans by and large weren't doing anything approaching what the English were doing in terms of audience appreciation. I think there was a tremendous sense that these shows were a step backwards from Prince and Sondheim's best work and the fact that they were more popular than any of our shows turned everyone into a sore loser. I personally didn't enjoy many of those shows but I will go to my grave not knowing whether that's resentment or taste.

How was it a backward step?

Well, it started with *Cats*. There were shows before that—there was *Jesus Christ Superstar* and *Evita* but they weren't hits on the level of *Cats* and they by and large had some American involvement. With *Cats*, first of all I don't think T.S. Eliot's cat poems have any cachet in the United States. Also I think there was a feeling (especially among admirers of Sondheim and that whole line of descent from Jerome Kern, George Gershwin, Irving Berlin, Richard Rodgers, and Cole Porter) that Lloyd Webber wasn't a composer in the same tradition. Whether it was true or not, that was what people felt.

I actually have a whole other theory about this. I'm also the artistic director of Encores! and it has occurred to me that there was a very clear lineage that ran from operetta in the teens to musical comedy to musical drama in the 1940s. They coexisted to some degree but there was a clear trend. I wonder whether what we really saw in the late twentieth century was the rise of the second generation of operetta, that with *The Producers* and *Hairspray* we were getting the second generation of musical comedies, and that we are now headed toward a second generation of musical dramas. *The New Moon*, which was done in 1928, feels very much like a Lloyd Webber show—it feels like *Phantom of the Opera*. *Les Mis* is like a *Desert Song* operetta.

There seems to be a tendency to lump all the London shows together.
Well, in part it's because they were mostly written and produced by one or two sets of people. There's an identifiable compositional and promotional style. And it can also be attributed to the marketing strategies. Cameron Macintosh invented those single-image posters. If you actually examine the material—the music, lyrics, and style of production—they may be dramatically different.

Was there a sense that the staging was terribly different from Broadway musicals or did things like the helicopter just turn everybody off?
I think people's reactions largely had to do with the spectacle and the fact that it was English. There are considerable staging differences. They're a little hard to quantify but there are stylistic tics that are popular in England in the same way that there are here. All that means is that there's a way of thinking about what a show should look like that's different there than it is here. It was interesting watching Nicholas Hytner's *Carousel*, for example. It didn't look anything like an American *Carousel* and it didn't deal with the way you get from one place to another like a typical American production. But in *Carousel* you don't have to. Other shows are very hard to do unless you use the ride-out in a certain way. They're written to be done that way.

Would you say that there is a legacy here in New York of the commercial British musicals?
Certainly as a producing and marketing issue. What Cameron and Andrew did drastically changed the way we all think about how to put on a show,

why you put on a show and who your audience is. They also discovered that the jet airplane has made it so possible for people to come to New York or London that the audience is a lot broader and a lot less sophisticated than it used to be. As a businessman, there is an advantage to creating a show for the broadest possible audience. They seized upon that in a way that none of us had.

Going back to your earlier theory, there does seem to be a sense of the current commercial musicals looking backwards.
It is going back, there's no doubt about it. The years they reference are never ours. I think the truth about that is that the light-hearted comedy world of contemporary America is not in the theatre at all. It's in stand-up, television, and movies. If we were doing something that was the equivalent of *Best Foot Forward* (which was a show in 1941 that was about boys getting caught in the girls' dorm on prom night) it would be a teen movie.

Everybody's making musicals from movies at the moment, which makes sense when you think about the fact that the kind of underlying material that people made popular musicals from in the 1950s was light novels and plays, which hardly exist any more.

The Prince musicals were so powerful partly because they were of the moment whereas now it seems to be all about distancing the material from the audience.
It's very hard to do and we talk about this a lot. We can't figure out how to do *How to Succeed in Business Without Really Trying*. We can't figure out how to do a musical in which people dress the way you and I are now dressed. We have to put a lens on it. Because if people walk out looking the way we look and start to sing everyone has a very hard time accepting that. And the only exception to that rule is *Rent*.

You have worked in nonprofit theatre and now in commercial theatre. What is your sense of the relationship between the two sectors in terms of musical theatre?
First of all, resident theatre grew closer to Broadway because all of Broadway's plays became transfers from resident theatres. And at a certain point it just seemed silly not to do it that way. What's the point of taking a risk of what is now $2—2.5 million on a play when you don't have to? You can do it some place else and find out whether you want to take that risk. There didn't seem to be any down-side to it. And from the resident theatre point of view, once they got over the notion that coworking wasn't necessarily bad there wasn't a down side either, because they could have the

productions enhanced and do them at a higher level. As long as they could keep the control issues straight (like who's the boss when the show's at the Mark Taper Forum as opposed to who's the boss when the show's on Broadway) there didn't seem to be a downside for anyone.

Musical theatre followed behind that. There were two hurdles, only one of which was overcome. When the first American Congress of Theatre happened about 30 years ago, the nonprofits and Broadway were bitter enemies—mainly over material, although the question of funding was also an issue. The nonprofit theatre had been set up as an antidote to Broadway and Broadway was symbolized by the musical. No one wanted an antidote to Arthur Miller—they wanted an antidote to Frank Loesser. So the first hurdle was the idea that musical theatre was a legitimate enterprise for non-profit theatre. Little by little people came to understand that music theatre, if not musical comedy, was a form of dramatic enterprise that should be explored. And then came the ancillary benefit that a commercial producer might donate a large amount of money to a resident theatre to develop a show that their audience would want to see. That was the lesson of *Big River*, which was developed largely with money donated by commercial producers to the La Jolla Playhouse, proved to be a big success for La Jolla, and then came to New York.

The hurdle that's never been overcome is the physical one. You can't do in any resident theatre what you can do at the highest level on Broadway in terms of scenery, costumes, scenic transitions, and special effects. It's not just a matter of money—the physical limitations of resident theatre buildings and shops won't allow it. The buildings weren't designed to accommodate the mission of doing big Broadway musicals with mecha-nization and fly galleries and so forth. So the shows that have been devel-oped in resident theatres by and large have been physically modest because, having conceived them for a theatre that has limitations, it's not worth reconceiving them for Broadway—you simply dress them up a little bit more. Events like *The Producers* and *Hairspray* get to be done less and less.

And there are no longer a lot of directors who know how to work in the old style. Jack O'Brien did it on *Hairspray* and Susan Stroman did it on *The Producers* but it's done less and less. It was very interesting watching Tommy Tune do *Grand Hotel* in Boston as one of the last people to come from the generation of directors who knows how to fix a show on the road: rehears-ing one version in the afternoon and playing another version at night, and then teching the new version the next afternoon and playing it that night. Directors from the nonprofit world often have no idea how to do that—and why would they want to try it? Their notion is that you do the show at a resident theatre, you look at it, you close it, you fix it up on paper and then you put it up again. It's not necessarily a negative evolution, although I do feel that old-fashioned showbiz is perpetually threatened by these kinds of developments. Now, whether old-fashioned showbiz ought to be allowed to die out quietly is another question. I don't think it should be, but I grew up on it.

The issue here seems to be that directors are increasingly coming out of the smaller nonprofit theatres rather than coming through the ranks of Broadway musicals.
Absolutely. The other thing is that you don't have the same kind of tide pool, to use the marine biology model. The way marine biology works is that you have tide pools with a thousand tiny fish. When the tide comes in and they go out to sea, 99 percent of them get eaten and a few survive—the strongest and smartest survive. In the days when you had 30 musicals a season on Broadway, you probably had 1,500 chorus dancers, some of whom would emerge as choreographers and directors. Now you have four musicals produced with small choruses, so you have a very small tide pool. And when they come to direct a musical, they don't come with the credentials they used to have. By the time that Peter Gennaro became a choreographer, he'd probably danced in 40 choruses and watched 15 different choreographers solve problems out of town. They came armed with a huge vocabulary of what to do when you got into trouble, what to do when you have a new idea, how long it's going to take to develop it—all that boiler-plate stuff. If you come up through a dance program and you've danced in a couple of shows, you don't really have that kind of background.

Do you think that the creative baton for creating new musicals has passed to the nonprofit sector or do you see it as a dialogue?
It's much more of a dialogue than it is with straight plays. With plays I think we essentially cede the entire territory to the nonprofits and we just shop, which is sort of a disgraceful state of affairs but that's what it is. With the musical I think there's much more collaboration, much more room for commercial producers to come up with ideas. There's a lot of music theatre that has no commercial possibilities at all that's being developed entirely by the nonprofit theatre companies. But simultaneously there's lots of activity where commercial producers and artistic directors can talk about shows.

The strongest position we [the commercial producers] could be in would be to bring the creative team (and therefore the script created by the team) and an enhancement of say $800,000. We would then say that we would like to control the production because we're bringing everything and they're bringing the facility. And they would probably say: "We have to work out a way to make that happen without us feeling like a rental house. We have an artistic mission, we have an artistic director and we have a dramaturg." So we would go back and forth to see whether we thought it was a good match or not. Barry Grove [Executive Producer of Manhattan Theatre Club] said to me once: "I love it when someone calls and says 'we'd love to do a show with your theatre.' I hate it when someone calls and says 'we'd like to do a show *at* your theatre.'" So to some degree it's a diplomatic matter, but there are other issues at play from casting right down to what the poster looks like.

How do you feel about the criticism that enhancement deals threaten the artistic integrity of nonprofit theatres?
It seems to me that companies that are more radical about what should be done on their stage won't touch this stuff. I mean, La MaMa's not going to do *Big River*. They have a different mission and they keep to it. Now, are individual theatres sometimes corrupted away from their vision by promises of cash and back-end royalty deals on a musical? Probably, and that's probably not good. But to some extent I think that if the artistic director and the managing director can't keep to their mission then that's their problem.

Turning it around, do you think that enhancement deals have helped to channel commercial money into shows that they might not otherwise have been able or willing to fund?
Usually when a commercial producer enhances a show it's because it has a strong commercial possibility. But I do think that the money thrown off by shows like *A Chorus Line* and *Bring in Da Noise, Bring in Da Funk* at the Public have helped to make things like *Caroline, or Change* possible—which is clearly not what you would initially think of as a Broadway show. Everyone in the theatre complains about how homogenizing Broadway is and to an extent I sympathize. But it's so much more diverse than when I was a high school kid. Many of the most interesting musicals on Broadway have come out of the notion that when commercial producers throw money at nonprofits, they get something back in the way of a more interesting show two years down the road. It's definitely producing some kinds of dividends in terms of pushing the envelope.

I am very interested in the recent surge of revivals that take a fresh approach to classic musicals—a trend that seems to be led by British directors.
It *has* been mainly British directors who have come in and done these newly conceived productions, some of which have worked and some of which

haven't. I think probably they're liberated to do these productions by the fact that they don't have any stake or ownership in what was done in the past. I think there are shows that are susceptible to being done well that way and shows that really aren't—they're constructed in such a way that you're better off just doing them as originally done.

It seems to me that some of the revivals have felt almost Chekhovian.
The problem there is that Chekhov is largely about repressed energy whereas musicals are largely about unrepressed energy. You can't write a musical (as far as I can tell) about a passive hero. You can write a musical about a hero who's a jerk (look at *My Fair Lady*) but he has to want something—and to want it so badly that he wants to open up his mouth and sing about it right away. That's not true of plays. You can do a play where a guy doesn't know how to express himself. And you can easily do a movie about a guy who's walking down the street and something happens to him. But in a musical, somebody has to want something or the game's not worth playing.

INTERVIEW WITH RICHARD MALTBY, JR.
(May 7, 2004, New York City)

Richard Maltby, Jr. is a bookwriter, lyricist, and director best known for Broadway shows like Ain't Misbehavin' *and* big. *As the co-lyricist of* Song and Dance *and* Miss Saigon, *Maltby was one of a handful of Americans artists to work on the West End musicals in the 1980s and 90s, giving him a rare trans-Atlantic perspective on this period.*

I am very interested by the fact that you went to London to work on *Miss Saigon*.
I describe myself as the Benedict Arnold of musical theatre. I was one of the only Americans who worked on the British Invasion musicals.

Why do you think there was such a strong reaction when London assumed the creative mantle for commercial musicals?
Americans did it best and British musicals were twee, second-rate, primitive imitations of American musicals. And then along came a couple of extraordinary big voices. And it happened to come at a time when American musicals had stopped being really inventive.

Andrew's two big ideas were that the book stopped musicals and that big melodies were the heartbeat of musical theatre in the biggest sense of the term, including opera. American musicals used to have the big melodic songs, but in the 1960s, American writers became increasingly subservient to the story. With the rise of rock music, they no longer attempted to write hit songs. Musicals had elaborate, book-oriented ideas. Some of them were quite wonderful and quite thrilling, but the idea of just standing there and thrilling you with a song that had a big melody had fallen into disuse. So when Andrew came along with the big melodies, it was startling and thrilling. Andrew's god is Puccini and later Richard Rodgers for the same reason—the big melody. It was thrilling for the audience. It wasn't thrilling for the writers and creators of American musicals who sat there and thought, "What on earth is *Cats* anyway? It's this bizarre kind of revue with some big songs. It's not anything that we know of. It's a collection of set-piece songs for different characters in a setting in which everyone is a cat." But because it broke all the rules and real theatrical energy, it was very exciting.

Andrew's impulse to eliminate the book led eventually to recitative, which was not the original impulse. The impulse originally was to do it all within song—songs that had the story within it. In *Joseph*, there was some recitative but it had melodies and only served to get to the next hit tune. Of course, opera also started to get more recitative and fewer hit tunes and pretty soon they had no hit tunes and just this endless wallpaper of singing. I think it was the death of opera, actually—the inability to write tunes.

Lloyd Webber's work in particular seems to have been attacked for pandering too much to mass audiences.
Musical theatre has always been vulgar—it is of the people. It has a lot of art in terms of the creation but the idea is to go for the jugular. Verdi, Puccini—they really want to knock you to your knees with fury. The other side of this is that it can be cheap, but cheap isn't necessarily bad if it's new. Cheap is something that really gets you, that comes straight at you and trots out the tricks. The skill is just hiding the cheap. Take away the cheap from the musical theatre canon and you haven't got much left.

With *Phantom* Andrew crossed into book musicals and there he's mired because he is sunk by the weakness of the stories and plotting. Early on he had risen above it.

Les Misérables and Miss Saigon seem to me very European as well as more operatic in sensibility, largely due to French composer Claude-Michel Schönberg.
It was the first time since the old operetta writers that the European sensibility came into the music. *Les Misérables* has a solid mass of songs, its recitative is very tuneful, and it always serves the purpose of moving the story forward. In *Miss Saigon*, he uses a common opera technique that is never done in musicals: in songs like *Last Night of the World* and *I Still*

Believe he reaches the sung climax and then the orchestra goes on for another 30 seconds. The singers are not singing on the last note of the music. Usually in musicals, the thrill is having the voice cutting off; here, the thrill is the sweep of the orchestra. Luckily he had an opera director who knew how to work with that. Nicholas Hytner had had all the training of European directors—Shakespeare, classics, opera—and he understood music.

I read that Cameron Macintosh brought you in to provide an American sensibility.
I think he was more right than he knew because the version of the show that came to me was closer to *Madam Butterfly*. Pinkerton the American was a complete shit. He just didn't care about this girl at all and more than that, none of them really understood the impact of Vietnam on the American psyche. It was an attitude of "we've been losing colonies for years—get over it! What's the big deal?" There was this American myth that we don't lose wars, that we always win, that John Wayne would come over the hill and save the day. The city of Saigon was surrounded, we were out and the soldiers that were there still thought that something would happen because it was unthinkable that it wouldn't. And when it suddenly became clear that it wasn't going to happen, it was jaw-droppingly horrendous. So my contribution was to give them an American sensibility. And also, unlike *Madam Butterfly* (where there's no impediment to the love story except that Pinkerton couldn't care less) here, because the city and the country completely closed down, you could really have an actual love story. There would be no way that they could get back to camp and, therefore, home to start a new life. So we had a story with no villains but with the metaphor implicit in this dopey American trying to help the dopey oriental girl. But he doesn't understand her world. They're in love but there's no way that he could ever understand the sensibility of the girl. I love the plot of the story because there are no villains.

Did you see the original version of the show as a concert spectacular in Paris?
Yes I saw it. It was like a pageant with music. They did the opening scene and they would sing. Then it would stop and you would sit there for five minutes while they changed the scenery. Then they'd do the next big moment and then stop again and change the scenery.

How crucial was Hytner in the development of *Miss Saigon*?
I can't think of another director who could have done the job that he did. He was new to it, he was primed for it, he hadn't unleashed everything yet

and this was the show that did it. If you read the script of *Miss Saigon*, it reads like a play with four of five big set pieces. I used to joke, what are all those people from the big numbers going to be doing for the rest of the evening? Little did I know that Nick would put the whole city onstage and that even while a scene in a room is happening, people are going by outside and there was this tremendous sense of the life of the city. It was he and Bob Avian who took "The American Dream" and decided to make it this dream sequence, which is a climactic moment. Underlying everything in the story is this idea that somewhere in America is the answer, the solution to all of this—so much so that Kim gives her life to make sure that her child has it. So "The American Dream" was both a production number and a very satisfying concluding thought.

How do you feel about labels like "technomusicals" that American theatre historians have attached to *Miss Saigon*?
It's the perfect example of Americans who didn't even look at the stage. It's the worst kind of chauvinism. Luckily the common folk didn't have any trouble with seeing it in terms of story and character. I saw a version of it in Copenhagen in the round with a much simplified production—no helicopter, smaller cast. It will play better and better the simpler it is. As you strip all these things away people will be astonished to find that it's a really strong show. Everyone's life is at stake all the way through.

How involved was Hytner with the writing process?
His main contribution or intervention was with the sacred stuff. Vietnamese girls had a mystical kind of religion that's in "The Sacred Bird." There was a lot of that and Nick thought that that just sounded like operatic mumbo-jumbo even though it was literally the truth of the characters. So he encouraged us to take out most of that. In a theatrical sense she would have come over as vaguely demented so we trimmed it down to mere ancestor worship. But generally the story didn't change. He took that and really dug into the scenes. He gave the ambience, the big story within which the little story was taking place. In the middle of "Why God, why?" Chris comes out of the house and is accosted by all the Vietnamese who, as soon as they see an American soldier, say "Get me out of here, get me out of here. Can you help?" And he pushes them away and goes back inside. That addition is completely telling. For "I Still Believe," someone told Hytner it was odd that this poor girl had a whole house to herself, and he restaged it with 20 people sleeping all over the floor and the Engineer had to find his way through them to Kim. It's a more realistic idea of what her life and the poverty were like. Everything that he did fed into the centre.

You have been quoted as saying that London in the early 1990s was like New York in the 1940s and 50s. Could you expand on that?

Great American musicals come in bursts. In the late 1940s when I was growing up, there were all these articles in the *New York Times* about the integrated musicals with Frank Loesser writing "you see, the songs will tell the stories." And it seemed there was something like that going on in England—that there was a discovery of a different kind of scale of singing and a different kind of show, the impact on an audience.

Notes ❧

INTRODUCTION

1. Anthony Tommasini, "They Do Write 'Em Like They Used To," *New York Times*, May 20, 2001.
2. Frank Rich, "The Empire Strikes Back," *New York Times*, March 29, 1987.
3. Richard Maltby Jr., interview with the author on May 7, 2004.
4. Notable exceptions include two books of interviews: Lawrence Thelen's *The Show Makers* (2002) and Jackson R. Bryer and Richard A. Davidson (eds.) *The Art of the American Musical: Conversations with the Creators* (2005), which include interviews with James Lapine and with George C. Wolfe, respectively. Jessica Sternberg's *The Megamusical* (2006) also includes some comments on directors although her primary focus is on the music and, secondarily, the producers.
5. Jessica Sternberg, *The Megamusical*, 5.
6. Ibid., 1, 5.
7. Aaron Frankel, *Writing the Broadway Musical*, 1–4.
8. Martin Gottfried, *More Broadway Musicals—Since 1980*, 28.
9. William Goldman, *The Season*, 285–298.
10. One study that does focus on directors is Lawrence Thelen's *The Show Makers*.
11. Later, Gertrude Lawrence would also exert an influence on the musical play: it was at her instigation that Rodgers and Hammerstein wrote *The King and I* as a vehicle for her talents.
12. For further background on this era in British musicals see Sheridan Morley, *Spread a Little Happiness*, 15–58.
13. For analysis of the central role of musical theatre in American culture see Ann Douglas, *Terrible Honesty—Mongrel Manhattan in the 1920s*, David Walsh and Len Platt, *Musical Theater and American Culture*, and John Bush Jones, *Our Musicals, Ourselves: A Social History of the American Musical Theater*.
14. John Degen, "Musical Theatre since World War II," in Don B. Wilmeth and Christopher Bigsby (eds.) *The Cambridge History of American Theatre*, 440–441.
15. Todd Gitlin, *The Sixties: Years of Hope, Days of Rage*, 16.
16. Savran, *A Queer Sort of Materialism*, 3.
17. Ibid.

18. Michael Kantor and Laurence Maslon, *Broadway: The American Musical*, 246–247.
19. Initially called "The Toast of the Town," the title was changed in 1955.
20. For further discussion see John Bush Jones, *Our Musicals, Ourselves*, 123–201.
21. Len Platt, *Musical Comedy on the West End Stage*, 128–129.
22. Richard Eyre and Nicholas Wright, *Changing Stages: A View of British Theatre in the Twentieth Century*, 134.
23. John Snelson, "'We Said We Wouldn't Look Back': British Musical Theatre, 1935–1960," in William A. Everett and Paul R. Laird (eds.) *The Cambridge Companion to the Musical*, 108–112.
24. Ibid., 107.
25. Noel Coward, quoted in Morley, 157.
26. Aronson, "American Theatre in Context: 1945–Present," in Don B. Wilmeth and Christopher Bigsby (eds.) *The Cambridge History of American Theatre*, 100.

1. HAROLD PRINCE IN CONTEXT

1. Goldman, *The Season*, 285–298.
2. Gitlin, *The Sixties*, 144–145.
3. These are the dates in which the theatres went professional. For further discussion see Berkowitz, *New Broadways*, 57–60.
4. Aronson, "American Theatre in Context: 1945–Present," in Don B. Wilmeth and Christopher Bigsby (eds.) *The Cambridge Companion to American Theatre*, 143.
5. Berkowitz, 117–118.
6. Barbara Means Fraser, "The Dream Shattered: America's Seventies Musicals," *Journal of American Culture*, 31.
7. Harold Prince, *Contradictions: Notes on Twenty-Six Years in the Theatre*, 68.
8. Martin Gottfried, *A Theater Divided*.
9. Max Wilk, *OK! The Story of Oklahoma!*, 222.
10. Wilella Waldorf, "Two on the Aisle," *New York Evening Post*, April 1, 1943.
11. Stephen Sondheim and George Furth, *Company*, 116.
12. Stephen Sondheim in an interview at Lincoln Center, June 2, 1975. Quoted in Foster Hirsch, *Harold Prince and the American Musical Theatre*, 21.
13. Prince, *Contradictions*, 37.
14. Ibid., 125–126.
15. Ilson, *Harold Prince: A Director's Journey*, 11.
16. Mark Steyn, "The Man Who Gets Everyone Dancing," *Daily Telegraph*, July 10, 1996.
17. Prince, *Contradictions*, 2.
18. "From Follies to Phantom," A & E Cable TV Network. Quoted in Ilson, *Harold Prince*, 11.

19. Andrea Most, *Making Americans*, 9.
20. Bertolt Brecht, "The Modern Theatre Is the Epic Theatre," in John Willett (ed. and trans.) *Brecht on Theatre: The Development of an Aesthetic*, 37.
21. Prince, *Contradictions*, 144.
22. "Side by Side by Side," *American Theatre* (19, no. 6, 2002) 69.
23. Sondheim and Furth, *Company*, 32.
24. Ibid., 48.
25. Craig Zadan, *Sondheim & Co*, 152–153.

2. FROM *CABARET* TO *SWEENEY TODD*: MUSICAL DRAMA ON BROADWAY

1. Kander, Ebb, and Laurence, *Colored Lights*, 60–61.
2. Prince, *Contradictions*, 137.
3. Ibid., 68.
4. Ibid., 7.
5. Kander, Ebb, and Laurence, *Colored Lights*, 60–63.
6. Joe Masteroff, John Kander, and Fred Ebb, *Cabaret: The Illustrated Book and Lyrics*, 19.
7. Alexander Gershkovich (trans. Micael Yurieff), *The Theater of Yuri Lyubimov*, 57.
8. Ibid., 58.
9. Prince, *Contradictions*, 129–130.
10. Ibid., 128.
11. Ibid., 126
12. Kander, Ebb, and Laurence, *Colored Lights*, 63.
13. See Ilson, *Harold Prince*, 130.
14. Gottfried, *Broadway Musicals*, 127.
15. Ilson, *Harold Prince*, 159.
16. Stanley Kauffman, "Company," *The New Republic*, May 23, 1970.
17. Prince, *Contradictions*, 73–74.
18. Brendan Behan, *The Hostage*, 170.
19. Stanley Kauffman, *Saturday Review*, November 24, 1979.
20. See Ilson, *Harold Prince*, 272.
21. Joan Goodman and Mike Bygrave, "Presenting the Che and Eva Show," *The Observer*, June 11, 1978.
22. Michael Coveney, *Cats on a Chandelier: The Andrew Lloyd Webber Story*, 80.
23. *Daily Telegraph*, June 23, 1978.
24. Coveney, *Cats on a Chandelier*, 81.
25. Goodman and Mike Bygrave, "Presenting the Che and Eva Show."
26. Ibid.
27. Walter Kerr, "'Evita,' a Musical Peron," *New York Times*, September 26, 1979.
28. See Hirsch, *Harold Prince and the American Musical Theatre*, 159.

29. Ilson, *Harold Prince*, 267; Hirsch, *Harold Prince and the American Musical Theater*, 165–166.
30. Dudley Andrew, "Echoes of Art: The Distant Sounds of Orson Welles," 173.
31. Hirsch, *Harold Prince and the American Musical Theater*, 166.
32. Joseph McBride, *Orson Welles*, 35.
33. James Naremore, *The Magic World of Orson Welles*, 63.
34. Hirsch, *Harold Prince and the American Musical Theatre*, 120.
35. Ibid.
36. The English translations of these titles are from Richard J. Hand and Michael Wilson, *Grand-Guignol: The French Theatre of Horror*.
37. Richard Eder, "Introducing Sweeney Todd," *New York Times*, March 2, 1979.
38. T.E.Kalem, "Razor's Edge: Sweeney Todd," *Time*, March 12, 1979.
39. Brecht, "The Modern Theatre Is the Epic Theatre," 36.
40. Bill Zakieriasen, "*Sweeney* Returns Opera to Its Roots," *Daily News*, June 4, 1979.

3. CULTURAL BARRICADES: READING THE WEST END MUSICALS

1. Aronson, "American Theatre in Context: 1945–Present," in Don B. Wilmeth and Christopher Bigsby (ed.) *The Cambridge History of American Theatre*, 138.
2. Savran, *A Queer Sort of Materialism: Recontextualizing American Theater*, 51.
3. Clive Barnes, "Come on Along and Listen to the Lullaby of High-Tech," *Sunday Times*, December 18, 1986.
4. Laurence Maslon, "A Changing Theatre: Broadway to the Regions—Broadway," in Don B. Wilmeth and Christopher Bigsby (eds.) *The Cambridge History of American Theatre*, 191.
5. The other two are the Shubert and the Nederlander organizations.
6. Jack Viertel, interview with the author on April 5, 2004.
7. Howard Kissell, "Viet Numb!" *Daily News*, April 12, 1991.
8. Edwin Wilson, *Wall Street Journal*, April 12, 1991.
9. Clive Barnes, "Smashing!" *New York Post*, March 13, 1987.
10. Michael Feingold, "Heat-Seeking Bomb," *Village Voice*, April 23, 1991.
11. Rosenberg and Harburg, *The Broadway Musical: Collaboration in Commerce and Art*, 7.
12. Singer, *Ever After: The Last Years of Musical Theater and Beyond*, 45.
13. Jones, *Our Musicals, Ourselves*, 322–324. Jones's justification for not classifying *Les Misérables* under his umbrella term "technomusical" is that the set "elegantly serves the visual requirements of the musical as one of many collaborating elements that together communicate a plot and characters of engaging and moving complexity."
14. Ibid., 325.

15. Clive Barnes, "A Stunning Evita Seduces with Its Gloss," *New York Post*, September 26, 1979.
16. Ilson, *Harold Prince*, 271.
17. Ibid., 270.
18. Clive Barnes, "Smashing!" *New York Post*, March 13, 2004.
19. Swain, "Operatic Conventions and the American Musical," in DiGaetani and Sirefman (eds.) *Opera and the Golden West*, 300.
20. Jack Viertel, interview with the author on April 5, 2004.
21. Steve Race, quoted in Coveney, *Cats on a Chandelier*, 203.
22. Brooks Atkinson, quoted in Alan Woll, *Black Musical Theatre: From Coontown to Dreamgirls*, 171. In fact, Porgy and Bess went on to receive productions in both theatres and opera houses, as did Prince-Sondheim collaborations *Sweeney Todd* and *A Little Night Music*.
23. Coveney, *Cats on a Chandelier*, 202–203.
24. Ibid., 202.
25. Bernard Levin, "The Cracked Mirror of Our Times," *Sunday Times*, July 30, 1978.
26. While the original leading man of *South Pacific* was opera singer Ezio Pinza, the rest of the cast was drawn from the Broadway world, including leading lady Mary Martin.
27. Saunders, quoted in John Story, "'Expecting Rain?' Opera as Popular Culture," in Jim Collins (ed.) *High Pop: Making Culture into Popular Entertainment*, 42.
28. Frank Rich, "'Miss Saigon' Arrives from the Old School," *New York Times*, April 12, 1991.
29. Frank Rich, "'Misérables,' Musical Version Opens on Broadway," *New York Times*, March 13, 1987.
30. Frank Rich, *New York Times*, April 12, 1991.
31. Robert Cushman, "Missing Link between Coward and Lloyd Webber," *The Telegraph*, May 1, 1999.
32. Michael Billington, "The Case for Irony in the Soul," *The Guardian*, December 5, 1983.
33. David Leveaux, interview with the author on June 20, 2003.

4. BEYOND THE LOGOS: WEST END MUSICAL DRAMA

1. Richard Maltby Jr., interview with the author on May 7, 2004.
2. Michael Billington, *Les Miserables*, October 10, 1985.
3. Trevor Nunn, "A Popular Front Steals the Show," *The Guardian*, October 4, 1986.
4. Ibid.
5. Frank Rich, "'Misérables,' Musical Version Opens on Broadway," *New York Times*, March 13, 1987.

6. John Caird, interview with the author on January 9, 2003.
7. A musical term referring to using only five tones—usually the first, second, third, fifth, and sixth tones of a diatonic scale.
8. Prece and Everett, "The Megamusical and Beyond: The Creation, Internationalization and Impact of a Genre," in William A. Everett and Paul R. Laird (eds.) *The Cambridge Companion to the Musical*, 247–248.
9. For a detailed account of the development process, see Edward Behr, *Les Misérables: History in the Making*.
10. Ibid., 67.
11. Robert Cushman, "The Naked and the Bed," *The Observer*, October 3, 1976.
12. John Caird, interview with the author on January 9, 2003.
13. Ibid.
14. Ibid.
15. Behr, *Les Misérables*, 98.
16. Ibid.
17. Morley, *Spread a Little Happiness*, 106.
18. Richard Maltby, Jr., interview with the author on May 7, 2004.
19. Ibid.
20. Edward Behr and Mark Steyn, *The Story of Miss Saigon*, 136.
21. Richard Maltby, Jr., interview with the author on May 7, 2004.
22. Behr and Steyn, *The Story of Miss Saigon*, 130.
23. Richard Maltby, Jr., interview with the author on May 7, 2004.
24. Ibid.
25. Behr and Steyn, *The Story of Miss Saigon*, 137.
26. Ibid.
27. Ibid.
28. Ibid.,133–136.
29. Alain Boublil and Claude-Michel Schönberg, *Miss Saigon: Easy Piano Solo/Vocal Album*, 65–66.
30. Bertolt Brecht, "The Rise and Fall of the City of Mahagonny," 233.
31. Behr and Steyn, *The Story of Miss Saigon*, 150.
32. Boublil and Schönberg, *Miss Saigon*, 64.

5. NEW HORIZONS: NONPROFIT MUSICAL DRAMA

1. Prece and Everett, "The Megamusical and Beyond: The Creation, Internationalization and Impact of a Genre," in William A. Everett and Paul R. Laird (eds.) *The Cambridge Companion to the Musical*, 265.
2. Steven Samuels, "Courage and Conviction," *American Theatre* (14, no. 4, 1997) 18.
3. Jaan Whitehead, "To Have and Have Not," *American Theatre* (18, no. 3, 2001) 25–26.

4. *Variety*, January 11, 1984 quoted in Rosenberg and Harburg, *The Broadway Musical*, 60.
5. Jack Viertel, interview with the author on April 5, 2004.
6. Bordman, *American Musical Theatre: A Chronicle*, 665.
7. The Billy Rose Theatre Collection, New York Public Library for the Performing Arts.
8. Ira Weitzman, interview with the author on July 30, 2003.
9. Hausam (ed.) *The New American Musical: An Anthology from the End of the Century*, xix.
10. James Lapine, interview with the author on September 23, 2003.
11. Ibid.
12. The first show, *In Trousers*, was produced earlier at Playwrights Horizons but, while functioning as a prequel to the other two shows, it is the latter two that made it to Broadway and that are usually referred to as the *Falsettos* musicals.
13. Ira Weitzman, interview with the author on July 30, 2003.
14. Ibid.
15. Ibid.
16. Lawrence Thelen's *The Show Makers* does include an interview with Lapine that makes reference to *Falsettos* in a more general discussion of his work.
17. William Finn, interview with the author on May 29, 2003.
18. Jones, *Our Musicals, Ourselves*, 336. Finn comment taken from his interview with the author on May 29, 2003.
19. Ira Weitzman, interview with the author on July 30, 2003.
20. Ibid.
21. Michael Feingold, "Changing Tenor," *Village Voice*, July 10, 1990.
22. Lyrics quoted from the liner notes of the original cast album.
23. James Lapine, interview with the author on September 23, 2003.
24. William Finn, interview with the author on May 29, 2003.
25. Ibid.
26. Ira Weitzman, interview with the author on July 30, 2003.
27. Ibid.
28. Frank Rich, "A Musical Theatre Breakthrough," *New York Times*, October 21, 1984.
29. James Lapine, interview with the author on September 23, 2003.
30. Stephen Sondheim in "Side by Side by Side," 68.

6. NONPROFIT DIRECTORS IN THE 1990s

1. Alex Witchel, "The Man Who Would Be Papp," *New York Times Magazine*, November 8, 1998.
2. Ed Morales, "Theatre and the Wolfe," *American Theatre* (11, no. 10, 1994) 20.
3. Marty Bell, *Broadway Stories: A Backstage Journey through Musical Theatre*, 49–50.

4. Morales, "Theatre and the Wolfe," 17.
5. As reprinted in the liner notes of the original cast album. Book and lyrics by Tom Eyen, music by Henry Krieger.
6. The production premiered in 1993 in Toronto and came to Broadway in 1994 via the West End.
7. George C. Wolfe, *The Colored Museum*, 24.
8. Ibid., 25.
9. Ibid., 30.
10. Ibid., 31.
11. Ibid., 32.
12. *Jelly's Last Jam*: book by Wolfe, lyrics by Susan Birkenhead, and music by Jelly Roll Morton. *Bring in Da Noise. Bring in Da Funk*: book by Reg E. Gaines, lyrics by Ann Duquesnay, Wolfe, and Reg E. Gaines, and music by Ann Duquesnay, Daryl Waters, and Zane Mark.
13. Margo Lion, interview with the author on April 2, 2004.
14. Ibid.
15. Ibid.
16. Ibid.
17. Rex Reed, *New York Observer*, May 4, 1992.
18. John Heilpern, *New York Observer*, May 4, 1992.
19. Frank Rich, *New York Times*, April 27, 1992.
20. Lyrics taken from the liner notes of the original cast album.
21. Margo Lion, interview with the author on April 2, 2004.
22. John Heilpern, "Breathtaking *Jelly's Last Jam* Is a Breakthrough Musical," *New York Observer*, May 4, 1992.
23. Frank Rich, *New York Times*, April 27, 1992.
24. Lyrics quoted from the liner notes of the original cast album.
25. Ibid.
26. Adam Guettel, interview with the author on April 9, 2003.
27. Ibid.
28. Tina Landau quoted in "When Is a Musical Not a Musical" in the Public Theater program of the subsequent Landau-Guettel collaboration, *Saturn Returns*, in 1998.
29. Quotation taken from John Guare's essay in the liner notes of the original cast album.
30. Adam Guettel, interview with the author on April 9, 2003.
31. Ibid.

7. RETHINKING REVIVALS

1. Amy S. Green, *The Revisionist Stage: American Directors Reinvent the Classics*, xi.
2. Ibid., 15.
3. Ted Chapin, *Everything Was Possible: The Birth of the Musical Follies*, 315.

4. Gower Champion died of a heart attack in 1980 followed within a few years by Bob Fosse (another heart attack) and Michael Bennett (AIDS). The Hal Prince-Stephen Sondheim partnership broke up in 1981 following the failure of *Merrily We Roll Along*. Jerome Robbins, while still active as a choreographer, retired from Broadway in the mid-1960s apart from staging revivals of his earlier work.
5. Jonathan Miller, *Subsequent Performances*, 23. Miller attributes the term to Aby Warburg's description of his Warburg Institute as a place where scholars could study "the afterlife of the antique."
6. Ibid., 23.
7. Ibid., 27–28.
8. Ibid., 55.
9. Ibid., 49.
10. Matthew Warchus, interview with the author on May 20, 2003.
11. Jack Viertel, interview with the author on April 5, 2004.
12. Panel discussion at Musical Theatre Works, New York City, May 14, 2002.
13. Todd Haimes, quoted in Charles Isherwood, "The New British Invasion," *The New York Times*, February 27, 2005.
14. Matt Wolf, *Stepping into Freedom: Sam Mendes at the Donmar*, 38.
15. David Leveaux, interview with the author on June 20, 2003.
16. Matthew Warchus, interview with the author on May 20, 2003.
17. Jack Viertel, interview with the author on April 5, 2004.

8. STAGING THE CANON: BRITISH DIRECTORS AND CLASSIC AMERICAN MUSICALS

1. Frank Rich, "You'll Always Walk Alone," *New York Times*, March 31, 1994.
2. David Richards, "A 'Carousel' for the 90's, Full of Grit and Passion," *New York Times*, March 25, 1994.
3. Ismene Brown, "Dancing with Joy," *The Daily Telegraph*, July 10, 1998.
4. *Variety*, March 28, 1994.
5. *National Theatre Platform Papers 4: Designers* (Royal National Theatre, 1993), 15.
6. Frank Rich, "You'll Always Walk Alone," *New York Times*, March 31, 1994.
7. Jane Edwardes, *Time Out*, February 16, 1992.
8. Frank Rich, "You'll Always Walk Alone," *New York Times*, March 31, 1994.
9. The production transferred to Broadway in 2002.
10. David Gritten, "The New Pioneer Spirit," *The Daily Telegraph*, July 10, 1998.
11. Andrea Most, *Making Americans*, 101–118.
12. Max Wilk, *OK! The Story of Oklahoma!*, 120.
13. Ismene Brown, "Dancing with Joy," *The Daily Telegraph*, July 10, 1998.
14. Ibid.
15. David Gritten, "The New Pioneer Spirit," *The Daily Telegraph*, July 10, 1998.

16. Michael Billington, "Oklahoma!" *The Guardian*, July 16, 1998.

17. Miller, *Subsequent Performances*, 70.

18. Sam Mendes quoted in Matt Wolf, "A London Maverick Arrives for First (and Second) Time," *New York Times*, February 1, 1998.

19. Walter Kerr, review of *Cabaret*, *New York Times*, November, 15, 1987.

20. Ben Brantley, "Desperate Dance at Oblivion's Brink," *New York Times*, March 20, 1998.

21. James E. Young, *The Texture of Memory: Holocaust Memorials and Meaning*, quoted in Mervyn Rothstein, "In Three Revivals, the Goose Stepping Is Louder," *New York Times*, March 8, 1998.

22. Mervyn Rothstein, "In Three Revivals, the Goose Stepping Is Louder," *New York Times*, March 8, 1998.

23. Michiko Kakutani, "Culture Zone; Window on the World," *The New York Times Magazine*, April 26, 1998.

24. Stephen Sondheim, quoted in Zadan, *Sondheim & Co*, 52–53.

25. Jack Viertel, interview with the author on April 5, 2004.

26. Zadan, *Sondheim & Co*, 167–170.

27. Mark Steyn, review of *Follies*, *The Independent*, July 23, 1987. There was a previous British production of the show at Manchester Library Theatre Company in 1985 but although this was nationally reviewed it was the commercial London production two years later that made a splash as the major British premiere.

28. Clive Barnes, "Revival's a Bit of a Folly," *New York Post*, April 6, 2001.

29. Ben Brantley, "Desperate Dance at Oblivion's Brink," *New York Times*, April 6, 2000.

30. Matthew Warchus, interview with the author on May 20, 2003.

31. Ibid.

32. Zadan, *Sondheim & Co*, 163.

33. Chapin, *Everything Was Possible*, 238.

34. Tommy Tune, *Footnotes*, 97.

35. David Leveaux, interview with the author on June 20, 2003.

36. Frank Rich, "*Nine*, a Musical Based on Fellini's 8 ½," *New York Times*, May 10, 1982.

37. David Leveaux, interview with the author on June 20, 2003.

38. The Broadway production was based on Leveaux's 1996 production of the show at the Donmar Warehouse in London.

39. David Leveaux, interview with the author on June 20, 2003.

40. Ibid.

41. Ibid.

42. Ibid.

9. THE LEGACY OF THE 1980s AND 90s

1. Barry Singer, *Ever After*, 209. Quotation from William Goldman, "Broadway Says Bye to 'Phantom' Funk," *Variety*, May 28, 2001.

2. This production first played in New York at the Avery Fisher Hall in Japanese. Subsequently, Miyamoto re-created the production for Roundabout Theatre Company in English using American performers.

3. Don Shewey, "Just What Is a Musical? Broadway Has a New Definition," *New York Times*, September 8, 2002.

4. Figures correct as of May 2007 as quoted on the NAMT website.

5. Zelda Fichandler, "The Profit in Nonprofit," *American Theatre* (17, no. 10, 2000) 30–33.

6. Jack Viertel, interview with the author on April 5, 2004.

7. Ibid.

Bibliography ❧

Andrew, Dudley. "Echoes of Art: The Distant Sounds of Orson Welles." *Perspectives on Orson Welles*. New York: G.K. Hall & Co, 1995.

Aronson, Arnold. "American Theatre in Context: 1945–Present." In Don B. Wilmeth and Christopher Bigsby (eds.) *The Cambridge History of American Theatre, Vol. III*. Cambridge; New York: Cambridge University Press, 2000.

———. *American Avant-Garde Theatre: A History*. London: Routledge, 2000.

Behan, Brendan. *The Complete Plays of Brendan Behan*. London: Eyre Methuen, 1978.

Behr, Edward. *Les Miserables: History in the Making*. London: Jonathan Cape, 1989.

Behr, Edward and Steyn, Mark. *The Story of Miss Saigon*. New York: Arcade Publishing, 1991.

Bell, Marty. *Broadway Stories: A Backstage Journey through Musical Theatre*. New York: Sue Katz & Associates, Inc., 1993.

Berkowitz, Gerald M. *New Broadways*. New York: Rowman and Littlefield, 1982.

Bigsby, Christopher. *Modern American Drama 1945–2000*. Cambridge; New York: Cambridge University Press, 2000.

Bordman, Gerald. *American Musical Theatre: A Chronicle*. Oxford; New York: Oxford University Press, 1992.

———. *American Operetta: From H.M.S. Pinafore to Sweeney Todd*. Oxford: Oxford University Press, 1981.

Boublil, Alain and Schönberg, Claude-Michel. *Miss Saigon: Easy Piano Solo/Vocal Album*. London: Wise Publications, 1991.

Brecht, Bertolt. "The Rise and Fall of the City of Mahagonny." In Bertolt Brecht, *Collected Plays 2*. London: Methuen Drama, 1994.

———. "The Modern Theatre Is the Epic Theatre." In John Willett (ed. and trans.) *Brecht on Theatre: The Development of an Aesthetic*. London: Methuen, 1993.

Bryer, Jackson R. and Davidson, Richard A. (eds.). *The Art of the American Musical: Conversations with the Creators*. New Jersey: Rutgers University Press, 2005.

Chapin, Ted. *Everything Was Possible: The Birth of the Musical Follies*. New York: Knopf, 2003.

Coveney, Michael. *Cats on a Chandelier: The Andrew Lloyd Webber Story*. London: Hutchinson, 1999.

Degen, John. "Musical Theatre since World War II." In Don B. Wilmeth and Christopher Bigsby (eds.) *The Cambridge History of American Theatre*. Cambridge; New York: Cambridge University Press, 2000.

Douglas, Ann. *Terrible Honesty—Mongrel Manhattan in the 1920s.* New York: Noonday Press, 1996.

Engel, Lehman. *The American Musical Theatre.* New York: Macmillan, 1975.

Eyre, Richard and Wight, Nicholas. *Changing Stages: A View of British Theatre in the Twentieth Century.* London: Bloomsbury, 2000.

Fichandler, Zelda. "The Profit in Nonprofit." *American Theatre* 17, no. 10 (2000): 30–33.

Frankel, Aaron. *Writing the Broadway Musical.* New York: Drama Books, 1977.

Fraser, Barbara Means, "The Dream Shattered: America's Seventies Musicals." *Journal of American Culture* 12, no. 2 (1989): 31–37.

Gershkovich, Alexander (trans. Micael Yurieff). *The Theater of Yuri Lyubimov.* New York: Paragon House, 1989.

Gitlin, Todd. *The Sixties: Years of Hope, Days of Rage.* New York: Bantam Books, 1993.

Goldman, William. *The Season.* New York: Limelight Editions, 1969.

Gottfried, Martin. *More Broadway Musicals—since 1980.* New York: Harry N. Abrams, Inc., 1991.

———. *Broadway Musicals.* New York: H.N. Abrams, 1979.

———. *A Theater Divided.* Boston: Little, Brown and Company, 1969.

Green, Amy S. *The Revisionist Stage: American Directors Reinvent the Classics.* Cambridge: Cambridge University Press, 1994.

Gussow, Mel. "Off and Off-Off Broadway." In Don B. Wilmeth and Christopher Bigsby (eds.) *Cambridge History of American Theatre Vol. III.* Cambridge; New York: Cambridge University Press, 2000.

Hand, Richard J. and Wilson, Michael. *Grand-Guignol: The French Theatre of Horror.* Exeter: University of Exeter Press, 2002.

Hausam, Wiley (ed.). *The New American Musical: An Anthology from the End of the Century.* New York: Theatre Communications Group, 2003.

Hillman, Jessica. "Goyim on the Roof: Embodying Authenticity in Leveaux's Fiddler on the Roof." *Studies in Musical Theatre, Vol. 1.1* (Intellect, 2007): 25–39.

Hirsch, Foster. *Harold Prince and the American Musical Theatre.* Cambridge; New York: Cambridge University Press, 1989.

Ilson, Carol. *Harold Prince: A Director's Journey.* New York: Limelight Editions, 2000.

Jones, John Bush. *Our Musicals, Ourselves: A Social History of the American Musical Theatre.* Hanover: Brandeis University Press, 2003.

Kander, John, Ebb, Fred, and Laurence, Greg. *Colored Lights.* New York: Faber and Faber, 2003.

Kantor, Michael and Maslon, Laurence. *Broadway: The American Musical.* New York: Bulfinch Press, 2004.

Koger, Alicia Kay. "Trends in Musical Theatre Scholarship: An Essay in Historiography." *New England Theatre Journal* 1 (1992): 69–85.

McBride, Joseph, *Orson Welles.* New York: Da Capo Press, 1996.

Maslon, Laurence. "A Changing Theatre: Broadway to the Regions—Broadway." In Don B. Wilmeth and Christopher Bigsby (eds.) *The Cambridge History of American Theatre, Vol. III*. Cambridge; New York: Cambridge University Press, 2000.

Masteroff, Joe, Kander, John, and Ebb, Fred. *Cabaret: The Illustrated Book and Lyrics*. New York: Newmarket Press, 1999.

Miller, Jonathan. *Subsequent Performances*. London: Faber and Faber, 1986.

Morales, Ed. "Theatre and the Wolfe." *American Theater* 11 (1994): 15–20.

Morley, Sheridan. *Spread a Little Happiness*. London: Thames and Hudson, 1987.

Most, Andrea. *Making Americans: Jews and the Broadway Musical*. Cambridge, MA: Harvard University Press, 2004.

Naremore, James. *The Magic World of Orson Welles*. New York: Oxford University Press, 1989.

Platt, Len. *Musical Comedy on the West End Stage*. New York: Palgrave Macmillan, 2004.

Prece, Paul and Everett, William A. "The Megamusical and Beyond: The Creation, Internationalization and Impact of a Genre." In William A. Everett and Paul R. Laird (eds.) *The Cambridge Companion to the Musical*. Cambridge; New York: Cambridge University Press, 2002.

Prince, Harold. *Contradictions: Notes on Twenty-Six Years in the Theatre*. New York: Dodd, Mead & Company, 1974.

Ratcliffe, Michael. Interview with Bob Crowley. *Platform Papers 4: Designers*. London: Royal National Theatre, 1993.

Rich, Frank. "Side By Side." *American Theatre* 19, no. 6 (2002): 20–24, 68–70.

Rosenberg Bernard and Harburg, Ernest. *The Broadway Musical: Collaboration in Commerce and Art*. New York: New York University Press, 1993.

Samuels, Steven. "Courage and Conviction." *American Theatre* 14, no. 4 (1997): 16–19.

Savran, David. *A Queer Sort of Materialism: Recontextualizing American Theater*. Ann Arbor: University of Michigan Press, 2003.

Singer, Barry. *Ever After: The Last Years of Musical Theatre and Beyond*. New York: Applause, 2004.

Snelson, John. "'We Said We Wouldn't Look Back': British Musical Theatre, 1935–1960." In William A. Everett and Paul R. Laird (eds.) *The Cambridge Companion to the Musical*. Cambridge; New York: Cambridge University Press, 2002.

Sondheim, Stephen and Furth, George. *Company*. New York: Theatre Communications Group, 1996.

Sternberg, Jessica. *The Megamusical*. Bloomington; Indianapolis: Indiana University Press, 2006.

Steyn, Mark. *Broadway Babies Say Goodnight: Musicals Then and Now*. London: Faber and Faber, 1997.

Story, John. "'Expecting Rain?': Opera as Popular Culture." In Jim Collins (ed.) *High Pop: Making Culture into Popular Entertainment*. Malden, MA: Blackwell, 2002.

Swain, Joseph P. "Operatic Conventions and the American Musical." In DiGaetani and Sirefman (eds.) *Opera and the Golden West*. Rutherford, NJ: Fairleigh Dickinson University Press, 1994.

Thelen, Lawrence. *The Show Makers*. New York: Routledge, 2002.

Tune, Tommy. *Footnotes*. New York: Simon and Shuster, 1997.

Walsh, David and Platt, Len. *Musical Theatre and American Culture*. Connecticut; London: Praeger, 2003.

Whitehead, Jaan. "To Have and Have Not." *American Theatre* (March 2001): 25–26.

Wilk, Max. *OK! The Story of Oklahoma!* New York: Grove Press, 1993.

Wolf, Matt. *Stepping into Freedom: Sam Mendes at the Donmar*. London: Nick Hern, 2002.

Wolfe, George C. *The Colored Museum*. New York: Grove Press, 1985.

Woll, Allen L. *Black Musical Theatre: From Coontown to Dreamgirls*. Baton Rouge: Louisiana State University Press, 1989.

Zadan, Craig. *Sondheim & Co*. New York: Harper & Rowe, 1986.

Index ∿